Molecular Industrial Mycology

MYCOLOGY SERIES

Edited by

Paul A. Lemke

*Department of Botany
and Microbiology
Auburn University
Auburn, Alabama*

Other Volumes in Preparation

Molecular Industrial Mycology
Systems and Applications for Filamentous Fungi

edited by

Sally A. Leong

Plant Disease Resistance Research Unit
USDA/ARS
Department of Plant Pathology
University of Wisconsin at Madison
Madison, Wisconsin

Randy M. Berka

Genencor International
South San Francisco, California

Marcel Dekker, Inc.   New York • Basel • Hong Kong

Library of Congress Cataloging-in-Publication Data

Molecular industrial mycology : systems and applications for
  filamentous fungi / edited by Sally A. Leong, Randy M. Berka.
      p.   cm. -- (Mycology series ; v. 8)
    Includes bibliographical references.
    Includes index.
    ISBN 0-8247-8392-1 (alk. paper)
    1. Fungi--Biotechnology.   I. Leong, Sally A.,
  II. Berka, Randy M.,   III. Series.
    [DNLM: 1. Fungi--genetics. 2. Genetic Engineering. 3. Industrial
  Microbiology.   4. Molecular Biology--methods. 5. Mycology.   QW 75
  M7183]
  TP248.27.F86M64   1991
  660'.62--dc20
  DNLM/DLC
  for Library of Congress   90-13979
                                                                        CIP

This book is printed on acid-free paper

MARCEL DEKKER, INC.
270 Madison Avenue, New York, New York 10016

Current printing (last digit):
10 9 8 7 6 5 4 3 2 1

PRINTED IN THE UNITED STATES OF AMERICA

Series Introduction

Mycology is the study of fungi, that vast assemblage of microorganisms which includes such things as molds, yeasts, and mushrooms. All of us in one way or another are influenced by fungi. Think of it for a moment—the good life without penicillin or a fine wine. Consider further the importance of fungi in the decomposition of wastes and the potential hazards of fungi as pathogens to plants and to humans. Yes, fungi are ubiquitous and important.

Mycologists study fungi either in nature or in the laboratory and at different experimental levels ranging from descriptive to molecular and from basic to applied. Since there are so many fungi and so many ways to study them, mycologists often find it difficult to communicate their results even to other mycologists, much less to other scientists or to society in general.

This series establishes a niche for publication of works dealing with all aspects of mycology. It is not intended to set the fungi apart, but rather to emphasize the study of fungi and of fungal processes as they relate to mankind and to science in general. Such a series of books is long overdue. It is broadly conceived as to scope and is to include textbooks and manuals as well as original and scholarly research works and monographs.

The scope of the series will be defined by, and hopefully will help define, progress in mycology.

Paul A. Lemke

Preface

Fungi have evolved to possess highly diverse and distinctive biological and biochemical systems that enable them to flourish in their specialized habitats. Besides the medical impact of these organisms, many fungi have exceptional industrial and economic significance. For example, toxin biosynthesis by plant pathogenic species of fungi can have a negative impact on agriculture. On the other hand, fungi have been utilized for decades in the commercial production of enzymes, antibiotics, and specialty chemicals. Thus, the complex biochemical and biological systems present in these organisms can have either positive or negative effects on society. Obviously, academic and industrial research communities are actively pursuing ways of minimizing the negative effects and maximizing the positive attributes of fungi as they influence all of us.

Molecular Industrial Mycology: Systems and Applications for Filamentous Fungi represents an overview of recent efforts by a selected group of researchers from a variety of disciplines who share a common interest—a desire to understand the genetics and molecular biology of these important systems in order to exploit them for the betterment of society. These researchers have provided us with up-to-date contributions from several key areas where fungi have an industrial or economic importance, including production of industrial enzymes, antibiotics, and toxins; heterologous protein secretion; and conversion of biomass. Research interest in these areas has greatly intensified in the last three years, primarily as a result of the application of modern molecular biology methods. Readers will undoubtedly notice that these powerful techniques are now being used to study virtually every organism discussed in this book. In comparison to genetically well-studied fungi such as *Neurospora crassa* or *Aspergillus nidulans*, the molecular genetics of most industrially important fungi is still in a stage of infancy. However, it seems likely that the knowledge which has accumulated from decades of research on well-characterized organisms will ultimately be applicable to the study of less-developed fungal systems. Additionally, as more fungi are analyzed, similarities are likely to be noted that will permit application of newly developed genetic technology across genus and species lines. As documented herein, this prediction has already been realized many times over.

We wish to communicate our gratitude to the contributors for their willingness to share their results and ideas. Finally, we express thanks to Carol S. Carol, Deborah J. Hope, and the rest of the staff of Marcel Dekker, Inc., for their assistance in assembling this volume.

Sally A. Leong
Randy M. Berka

Contents

Contributors

Andrawis Andrawis*   Department of Molecular and Cell Biology, The Pennsylvania State University, University Park, Pennsylvania

D. James Ballance   Delta Biotechnology Limited, Nottingham, England

R. Wayne Davies   Robertson Institute of Biotechnology, Department of Genetics, University of Glasgow, Glasgow, Scotland

Anu Harkki[†]   Biotechnical Laboratory, The Technical Research Center of Finland, Espoo, Finland

Erika L. F. Holzbaur[‡]   Department of Molecular and Cell Biology, The Pennsylvania State University, University Park, Pennsylvania

Thomas D. Ingolia   Department of Medical Chemistry, Lilly Research Laboratories, Indianapolis, Indiana

Kenneth Kaushansky   ZymoGenetics, Inc., Seattle, Washington

Jonathan Knowles[§]   Biotechnical Laboratory, The Technical Research Center of Finland, Espoo, Finland

Ashok A. Kumar   ZymoGenetics, Inc., Seattle, Washington

Current affiliations:
*Botany-Life Science Institute, Hebrew University, Jerusalem, Isreal
†Technology Center, Cultor Ltd., Kantvik, Finland
‡Worcester Foundation for Experimental Biology, Shrewsbury, Massachusetts
§ Glaxo Institute for Molecular Biology, Geneva, Switzerland

Juan F. Martin Department of Ecology, Genetics and Microbiology, University of León, León, Spain

Gary L. McKnight ZymoGenetics, Inc., Seattle, Washington

K. M. Helena Nevalainen Research Laboratories, Alko Ltd., Helsinki, Finland

Merja E. Penttillä Biotechnical Laboratory, The Technical Research Center of Finland, Espoo, Finland

Stephen W. Queener Lilly Research Laboratories, Indianapolis, Indiana

Nigel J. Smart* Pilot Plant Group, Allelix Biopharmaceuticals, Mississauga, Ontario, Canada

Tuula T. Teeri Biotechnical Laboratory, The Technical Research Center of Finland, Espoo, Finland

Sheryl A. Thompson Department of Molecular Biology, Genencor International, South San Francisco, California

Ming Tien Department of Molecular and Cell Biology, The Pennsylvania State University, University Park, Pennsylvania

Alan Upshall ZymoGenetics, Inc., Seattle, Washington

Jonathan D. Walton Department of Energy, Plant Research Laboratory, Michigan State University, East Lansing, Michigan

Michael Ward Research and Development, Genencor International, South San Francisco, California

*Current affiliation: Fermentations Operations, SmithKline Beecham Pharmaceuticals, King of Prussia, Pennsylvania

1

Transformation Systems for Filamentous Fungi and an Overview of Fungal Gene Structure

D. James Ballance
Delta Biotechnology Limited, Nottingham, England

I. INTRODUCTION

Filamentous fungi, and in particular *Aspergillus nidulans* and *Neurospora crassa*, have been extensively used as models for eukaryotic gene organization and gene regulation. These organisms are very amenable to such studies, and the methodology for classical genetic analysis is highly sophisticated, making use of sexual and parasexual crosses, as well as heterokaryotic and diploid states. The extreme metabolic versatility of filamentous fungi and their characteristic cell differentiation provide areas of study not possible in *Saccharomyces cerevisiae*.

Filamentous fungi have been employed, albeit unwittingly in many instances, in the production of fermented foods for many centuries. Exploitation of these organisms has expanded in recent times in controlled fermentations for the production of metabolites and enzymes. Current practices utilize genetically improved strains in which selection for desired characteristics, such as high yield and good fermentation properties, has resulted in highly productive strains, often working at close to the maximum theoretical capacity. Such strains have been obtained by the laborious means of mutagenesis followed by screening of survivors for the desired characteristics in a very empirical manner. Such procedures have, however, proved extremely successful and are still in use today.

The enormous impact of recombinant DNA methodology, and in particular transformation, on genetic analysis in *S. cerevisiae* provides ample demonstration of the value of such techniques (Struhl, 1983). Although the development of transformation methodology for filamentous fungi occurred more recently, the value is already evident. The ability to clone fungal genes and to reintroduce them into the organism allows the extension of classical genetic analysis to the molecular level.

Another benefit is the ability to introduce novel biosynthetic capacity and so exploit the efficient secretory mechanisms of filamentous fungi to produce economically important biologically active proteins (see Chapters 3 and 4). There is also scope to augment the classical mutation and selection approach for the improvement of industrial production strains by increasing gene dosage or by eliminating activities (see Chapter 2).

II. TRANSFORMATION OF FILAMENTOUS FUNGI

Transformation has now been reported for a wide range of filamentous fungal species (Table 1), and the general characteristics of these systems are very similar. Firstly, it is necessary to remove the permeability barrier represented by the cell wall either by treatment with lithium acetate (Dhawale et al., 1984) or, more commonly, by enzymic digestion with crude extracts of snail gut (Case et al., 1979) or a *Trichoderma* extract commercially available as Novozym 234 (Ballance et al., 1983) to produce protoplasts. Some species require other enzyme mixtures to produce sufficient quantities of regenerable protoplasts (Ballance and Turner, unpublished observations; Binninger et al., 1987). Removal of the cell wall requires that an osmotic stabilizer such as 0.6 KCl or 1.2 M sorbitol be present and, again, the most suitable stabilizer depends on the organism or on the final selection regime (Tilburn et al., 1983). DNA is added to the protoplast suspension in the presence of 10-50 mM $CaCl_2$ and then addition of a solution of polyethylene glycol (molecular weight 4000-8000) results in the uptake of the DNA. The treated protoplasts are then plated out on a medium that permits the regeneration of only cells that have acquired the transforming DNA. Alternatively, application of the selection pressure can be delayed until the protoplasts have regenerated cell walls, since they are particularly sensitive to antimetabolites prior to this (Ward et al., 1986).

Table 1 Transformed Filamentous Fungi

Species	References
Absidia glauca	Wöstemeyer et al. (1987)
Achlya ambisexualis	Manavathu et al. (1987)
Achlya ambisexualis	Cullen and Leong (1986)
Aphanocladium album	Daboussi et al. (1989)
Aspergillus nidulans	Ballance et al. (1983)
Aspergillus niger	Kelly and Hynes (1985)
Aspergillus oryzae	Mattern et al. (1987)
Aspergillus terreus	Upshall (1986a)
Beauveria bassiana	Daboussi et al. (1989)
Botryotinia squamosa	Huang et al. (1989)
Cephalosporium acremonium	Penalva et al. (1985)
Cochliobolus heterostrophus	Turgeon et al. (1987)
Colletotrichum graminicola	Turgeon et al. (1987)
Colletotrichum lindemuthianum	Daboussi et al. (1989)
Coprinus cinereus	Binninger et al. (1988)
Fulvia fulva	Oliver et al. (1987)
Fusarium oxysporium	Kistler and Benny (1988)
Fusarium solani	Marek et al. (1989)
Fusarium sporotrichiodes	Turgeon et al. (1987)
Gaeumannomyces graminis	Henson et al. (1988)
Gliocladium virens	Thomas and Kenerley (1989)
Glomerella cingulata	Turgeon et al. (1987)
Leptosphaeria maculans	Turgeon et al. (1987)
Magnaporthe grisea	Parsons et al. (1987)
Mucor circinelloides	van Heeswijk and Roncero (1984)
Nectria haemotococca	Turgeon et al. (1987)
Neurospora crassa	Case et al. (1979)
Penicillium caseicolum	Daboussi et al. (1989)
Penicillium chrysogenum	Beri and Turner (1987)
Penicillium nalgiovense	Geisen and Leistner (1989)
Podospora anserina	Begueret et al. (1984)
Schizophyllum commune	Munoz-Rivas et al. (1986)
Septoria nodorum	Coolley et al. (1988)
Trichoderma reesei	Penttila et al. (1987)
Ustilago hordei	Holden et al. (1988)
Ustilago maydis	Wang et al. (1988)
Ustilago nigra	Holden et al. (1988)

A. Selectable Markers

Most commonly used are genes that complement a nutritional mutation and permit growth in the absence of supplementation. The cloned gene need not necessarily be from the recipient fungal species, as there are many examples of interspecies and inter-genus gene expression (Kelly and Hynes, 1985; Ballance et al., 1983; Turgeon et al., 1985). One drawback of this approach is the requirement for a mutant recipient strain, and although selection of certain specific mutations is relatively easy, e.g., selection of mutants resistant to 5-fluoro-orotic acid for the generation of orotidine-5'-phosphate-decarboxylase-deficient mutants (Boeke et al., 1984; van Hartingsveldt et al., 1987), this is often undesirable, particularly in industrial strains. Fortunately there are a number of positive selection systems available based on resistance to antimetabolites, e.g., oligomycin resistance (Ward et al., 1986), though this resistance gene is probably species specific. Prokaryotic antibiotic resistance genes such as those for kanamycin or G418 resistance (Penalva et al., 1985) and hygromycin B resistance (Punt et al., 1987) have also been used, though the direction of transcription by a fungal promoter is required for efficient transformation. The latter requirement was exploited by Turgeon et al. (1987) to develop a system for the selection of sequences with promoter activity in *Cochliobolus heterostrophus*, though the procedure should be universally applicable.

A very useful alternative system uses the *A. nidulans amdS* gene encoding acetamidase, an enzyme that permits growth on acetamide as the sole carbon or nitrogen source (Kelly and Hynes, 1985). Even wild type *A. nidulans*, which already has one copy of the gene, can be transformed with the cloned *amdS* gene by virtue of enhanced growth on acetamide as the nitrogen source as a result of an increase in the gene copy number (Kelly and Hynes, 1985). Several other fungal species, including *A. niger*, appear to lack an acetamidase gene, so selection for transformants is even easier (Kelly and Hynes, 1985; Turgeon et al., 1985; Beri and Turner, 1987). The *A. nidulans amdS* gene functions in a wide range of filamentous fungi, making it a very useful marker for transformation. The frequency of transformation is generally 10-100 stable transformants/μg, though higher frequencies have been reported (Akins and Lambowitz, 1985; van Heeswijck and Roncero, 1984).

B. Fate of Transforming DNA and Stability of Transformants

Although the immediate fate of the transforming DNA on entry into the recipient cell is unknown, by the time a transformant has been purified, the DNA has, in almost all cases, integrated into the chromosomes by recombination (Case et al., 1979; Ballance et al., 1983; Tilburn et al., 1983). The three types of integration defined by Hinnen et al. (1978) are all found (Yelton et al., 1984), though the relative frequencies are dependent on the organism, strain, or even the marker used (Yelton et al., 1984; Case, 1986; Binninger et al., 1987; Wernars et al., 1986; Kim and Marzluf, 1988; Ballance and Turner, 1986). Usually, 30-50% of transformants have the plasmid integrated at the normal locus of the cloned gene, either by a single or double cross-over (or gene conversion) (Yelton et al., 1984). The former results in integration of the whole plasmid and can be extended into a tandem array of many copies. Integration at other sites must rely on very limited homology if, indeed, any is required. No systematic study of the minimum requirements for recombination in fungi has been undertaken, despite its obvious significance.

In general, transformants are mitotically stable, even in the absence of selective pressure, though they often show instability through the sexual cycle, particularly in *N. crassa* (Dhawale and Marzluf, 1985; Grant et al., 1984) (see below). Tandem arrays of integrated plasmids can undergo a reduction or amplification in copy number if the transformant is self-crossed (Kelly and Hynes, 1987), a characteristic that can be very useful in the study of the effects of gene dosage (Kelly and Hynes, 1987). There can also be a degree of instability manifest in parasexual analysis of transformants such that the transformation phenotype is not detectable in any of the haploid segregants, and this is apparently not due to preferential loss of the chromosome bearing the integrated plasmid (Ballance and Turner, 1985).

It is possible to recover the integrated transforming DNA from transformants by digestion with restriction enzymes, followed by circularization with T4 DNA ligase and then transformation of *E. coli* (Yelton et al., 1984; Ballance and Turner, 1985). By employing only partial restriction enzyme digestion, large areas flanking the inserted DNA can also be recovered (Ballance and Turner, 1985). Rather surprisingly, it is often possible to recover plasmids by transforming *E. coli* with untreated transformant DNA, even in the case of transformants in which the DNA has been unequivocally shown to have integrated into a chromosome (Ballance and Turner, 1985). It is assumed that there is a limited degree of "pop-out," or reversal of the integration event, leading to a very low proportion of free molecules that are not detectable on Southerns of DNA from stable transformants. There is no evidence that such molecules are able to replicate autonomously.

There have been numerous attempts to develop vectors that are able to replicate autonomously in filamentous fungi, the rationale being that the efficiency of transformation might be increased, as is the case in *S. cerevisiae* (Beggs, 1978). Replicating vectors might also facilitate the manipulation of copy number and recovery of the transforming DNA in *E. coli*. The latter is one of the main diagnostic criteria employed by many investigators (e.g., Stahl et al., 1987), even though it has been shown to occur with integrants (Ballance and Turner, 1985).

Another line of evidence for autonomous replication is instability of the transformant phenotype (Barnes and MacDonald, 1986), though this can be due to heterokaryosis (Upshall, 1986b). Meiotic instability of *Neurospora* transformants (Grant et al., 1984) seems likely to be related to methylation of the duplicated DNA, which leads to gross rearrangements in progeny from crosses (Selker et al., 1987).

In the light of the above outlined difficulties, most claims for autonomously replicating vectors should be treated with caution. Even the presence of apparently free plasmid molecules on Southerns can be explained as "popped-out" molecules, since integrated DNA is usually also present.

There are, however, two exceptions worthy of mention: van Heeswijck (1986) has shown conclusively that plasmids containing a fragment of *Mucor circinelloides* DNA able to complement a leucine auxotroph replicate autonomously when introduced into *M. circinelloides*. There was no evidence of chromosomal integration of the plasmids, and the transformants were mitotically unstable. Deletion of the *S. cerevisiae* and *E. coli* sequences from the plasmid did not affect their behavior, indicating that the sequences responsible for autonomous replication were within a 4.4-kb *Pst*I fragment of *Mucor* DNA. These replicating plasmids were present at a copy number of two to four copies per haploid genome.

Perrot et al. (1987) showed that *Tetrahymena thermophila* telomeric sequences permitted autonomous replication of plasmids as linear molecules in *Podospora anserina*. Approximately 50% of transformants had the plasmid chromosomally integrated but had lost the *T. thermophila* telomeric DNA. The remainder contained free linear molecules at less than one copy per nucleus and no integrated plasmid DNA and were mitotically unstable.

C. Gene Replacement

Development of techniques for gene replacement, whereby a cloned gene is used to replace its normal chromosomal copy by transformation, is of vital importance in the molecular dissection of gene function. It permits the evaluation of the effect of mutations introduced into the cloned gene at its normal genomic location, thereby eliminating spurious position effects. Miller et al. (1985) demonstrated that the one-step and two-step gene replacement methodology developed for *S. cerevisiae* were equally applicable to *A. nidulans*. However, difficulties have been encountered when trying to apply these techniques to *N. crassa* (Dhwale and Marzluf, 1985; Case, 1986).

A related technique of use both academically and industrially is to disrupt gene function by homologous recombination using a defective gene or one that is interrupted by a selectable marker (Miller et al., 1985). In this manner, null mutations can be created, and this can be used to confirm the cloning of a gene or to eliminate an undesirable gene activity in an industrial strain (see Chapter 2).

Transformation of diploids is a technique that has been employed extensively in *S. cerevisiae* for the disruption of essential genes. The lethality of the disruption is evident when the transformed diploid is allowed to undergo meiosis and sporulate, since half of the haploid products should carry the disrupted gene (Rothstein, 1983). Osmani et al. (1988) wished to investigate the effect of disrupting a cell-cycle control gene, *bimE*, in *A. nidulans*, but unfortunately the diploid approach is not feasible in this organism, since vegetative diploids do not undergo meiosis. Instead they developed a method using a haploid strain, utilizing the fact that protoplasts used for transformation tend to be multinucleate and can persist as heterokaryons in the coenocytic transformant colonies (Upshall, 1986b). However, since conidia are uninucleate, germination on nonselective medium should allow the growth of conidia derived from both nuclear types. The absence of the selectable marker (*pyr4*) in all germinated conidia indicated that the presence of this marker, and therefore gene disruption, was lethal.

III. GENE CLONING

A number of filamentous fungal genes have been cloned by the generally applicable techniques of differential hybridization (Hynes et al., 1983; Lockington et al., 1985), use of synthetic probes based on protein sequence data (Kinnaird et al., 1982; Boel et al., 1984), or use of antibodies raised against the purified protein (e.g., Viebrock et al., 1982; Shoemaker et al., 1983). Others have been isolated by expression in *E. coli* (Schweizer et al., 1981; Keesey and DeMoss, 1982; Yelton et al., 1983) or *S. cerevisiae* (Berse et al., 1983; Penttila et al., 1984), even though fungal promoters rarely function in these organisms and introns are inefficiently removed (see below). These two problems were circumvented by McKnight et al. (1985), who used a cDNA expression library to isolate the *A. nidulans* *adh3* and *tpiA* genes in yeast by complementation of the corresponding mutations (McKnight et al., 1985, 1986). Since the above approaches are limited in their application, much effort has been put into the development of methods for isolating genes by transformation of a filamentous fungus with DNA from gene libraries. Such systems enable the identification and cloning of any gene for which there is a discernible phenotype, be it relief of a mutation or a change in the appearance of the fungus.

The first report of gene cloning in filamentous fungi was by van Heeswijk and Roncero (1984) who transformed a leucine auxotroph of *Mucor circinelloides* with a gene library based on the *S. cerevisiae* vector YRp17. The cloned fragment of DNA fortuitously contained a sequence that promoted high-frequency transformation and autonomous replication (see above), and this may have facilitated cloning of the gene.

Most other systems rely on a lower frequency of transformation, though the effects of this can be reduced by using a gene library made in a cosmid vector, thereby reducing the number of clones to be screened.

In general, a gene libary is constructed in a plasmid or cosmid using established procedures of partial restriction with, for instance, *Sau*3A followed by ligation of size-fractionated fragments into the phosphatased vector (Maniatis et al., 1982). The vector usually contains a selectable marker to enable double selection for this marker and the gene of interest, thereby reducing the potential problem of spontaneous mutation. For most filamentous fungi, a cosmid library should contain in excess of 4000 clones to have a 99% chance of containing a given gene (Clarke and Carbon, 1976). However, great care must be exercised in the generation of genomic fragments for the library to ensure that it is truly representative.

In many instances it is possible to select for the gene of interest directly, for instance, when complementing a nutritional mutation (Ballance and Turner, 1986). In other cases it may be necessary to recover all transformants obtained by selection for the marker on the vector and subsequently screen for the gene of interest visually (Yelton et al., 1985) or by assay on a different medium.

A. Recovery of a Cloned Gene

With the above-mentioned exception of *Mucor*, a transformant with the appropriate phenotype will contain the selected DNA integrated into the genome. In the case of a cosmid library, it is possible to recover the cloned

DNA by in-vitro packaging of DNA from the transformant followed by infection of *E. coli* (Yelton et al., 1985). This procedure makes use of the lambda *cos* sites present in the vector, which are recognized by constituents of the packaging extracts. However, in cases where only one *cos* site is present, the DNA is still packaged efficiently, and this may be due to there being a small number of "popped-out" circular molecules, which would act as substrates for packaging.

When plasmids are used, the cloned DNA can be recovered by plasmid rescue, as outlined above. In some instances it is possible to recover "popped-out" molecules (Johnstone et al., 1985).

B. Sib Selection

Unfortunately, rescue of transforming DNA from *Neurospora crassa* appears to be very difficult, and this may prove to be the case for other filamentous fungi. To circumvent this problem, Akins and Lambowitz (1985) developed a procedure termed *sib selection*, which involved assembly of the gene library into pools of independent *E. coli* clones and transformation of *N. crassa* with DNA from these pools. When a pool was identified as containing the gene of interest, the clones making up this pool were subdivided into smaller pools, and so on until the strain was transformed with the DNA of the individual clones, thus identifying the clone containing the gene. Vollmer and Yanofsky (1986) foreshortened this procedure by employing a cosmid-based gene library, which facilitated storage of the library as independent clones and reduced the number of rounds of sib selection required to identify a clone containing the gene from five to three.

IV. FILAMENTOUS FUNGAL GENE STRUCTURE

A large number of genes of filamentous fungi have now been cloned and information on their primary structure is accumulating. The most extensively studied in this respect are the Euascomycetes *Neurospora crassa* and *Aspergillus nidulans*, though the general characteristics of filamentous fungal genes seem very similar. As is the case in *S. cerevisiae*, clustering of functionally related genes is rare, and when it is present there is no evidence of polycistronic messages, though the *aroM* genes of *N. crassa* and *A. nidulans* encode pentafunctional proteins (Charles et al., 1986) and the *trp-1/trpC* genes encode an enzyme with three activities (Schechtman and Yanofsky, 1983; Mullaney et al., 1985). Clustering of genes involved in sporulation in *A. nidulans* is quite prevalent (Zimmerman et al., 1980) and there is evidence of cluster-specific positional effects on the expression of some of these genes (Miller et al., 1987). The genomes of *A. nidulans* and *N. crassa* contain very little repetitive DNA (Timberlake, 1978; Krumlauf and Marzluf, 1980), and there is as yet no evidence for the existence of transposable elements in these organisms.

A. Initiation of Transcription

Initiation of transcription and its regulation have been extensively studied in many organisms, as gene expression is predominantly regulated at this level in most organisms, and this is also true for filamentous fungi. Sequences conserved in higher eukaryotic promoters, CAAT and TATA boxes (Benoist et al., 1980; Gannon et al., 1979), are sometimes evident in filamentous fungal 5'-non-transcribed regions (e.g., Kinnaird and Fincham, 1983; McKnight et al., 1985; Boel et al., 1984), but there are many instances where they are clearly absent (e.g., May et al., 1987; Kos et al., 1988), though functional equivalents may be present. When a sequence resembling the canonical TATA box ($TAT^A/_TAA/_T$) is present, the distance between this element and the transcription initiation points is less variable than in *S. cerevisiae* and is usually within 30-60 bp of the major start point. There are frequently a number of distinct transcription initiation points, often spread over some distance, but there is no evident consensus for the sequences at these start points. However, in many instances the sites of initiation are in regions that are rich in pyrimidines, this being particularly marked in the *A. nidulans oliC* gene (Ward and Turner, 1986). Preliminary deletion analysis of the *A. nidulans trpC* promoter has indicated that the TATA box and the pyrimidine-rich tracts may be important in positioning the sites of transcription initiation (Hamer and Timberlake, 1987).

There is much interest in the interaction of regulatory proteins with sequences within the promoters of genes that they control, and this has been particularly well studied in the quinic acid utilization (*qa*) cluster in *N. crassa* (Baum et al., 1987). In addition to the control of gene expression at the level of transcription initiation, it is clear that there are also position effects in the *spoC1* sporulation-specific gene cluster of *A. nidulans* (Miller et al., 1987).

B. Initiation of Translation

Kozak (1978) proposed a model for the initiation of translation in eukaryotes in which the ribosomal subunits scan the messenger RNA from the 5' end and initiate translation at the first AUG triplet encountered. However, a small proportion of proteins initiate at a downstream AUG (Kozak,

1983), so it is apparent that the context of the triplet is important, and indeed there is a high degree of conservation of the sequence around the initiator codon, $GCC^A/_GCCAUGG$ being the consensus in mammalian mRNAs (Kozak, 1987) and $^A/_YA^A/_UAAUGUCU$ in *S. cerevisiae* (Cigan and Donahue, 1987). The most important position would appear to be −3 (the A of AUG being +1 and the preceding base −1), at which a purine is nearly always present (97%) (Kozak, 1987), and usually as an A.

Table 2 shows that the AUG environment is also highly conserved in filamentous fungi, though there might be slight differences between *A. nidulans* and *N. crassa* upstream of −4. There is a preference for G at +4 and, more markedly, C at +5, which may reflect a tendency for proteins to have particular N-terminal amino acids rather than an importance in recognition by the ribosomal subunits.

Table 2 Frequency of Bases Around the Translation Initiation Codon

Organism	Base	\-10	\-9	\-8	\-7	\-6	\-5	\-4	\-3	\-2	\-1	+4	+5	+6
N. crassa	G	7	5	2	2	*12*	1	2	2	0	4	*18*	3	11
	A	11	11	6	12	*13*	8	2	30	*15*	18	3	6	1
	T	4	7	3	9	6	*17*	1	0	1	0	7	6	12
	C	9	8	20	10	2	8	29	2	*18*	12	4	*17*	7
A. nidulans	G	8	7	2	5	6	3	4	4	2	3	10	9	3
	A	5	9	8	*8*	6	8	6	26	*11*	15	5	5	5
	T	8	4	12	2	7	8	2	1	5	1	10	0	13
	C	8	9	7	*14*	10	13	20	1	*14*	13	6	*17*	7
All filamentous fungi	G	25	14	6	10	26	9	10	15	3	7	40	16	21
	A	24	37	20	27	24	19	19	*76*	*39*	50	16	18	8
	T	17	19	25	20	25	34	4	2	12	3	22	11	35
	C	24	21	40	36	18	37	66	6	*45*	39	16	*49*	24
Consensus (n = 99)							C	A	$^C/_A$	$^A/_C$			C	

Data were compiled from the sequences listed in Appendix 1.
The consnesus sequence was assigned according to the following criteria: If the frequency of a single nucleotide is greater than or equal to 50% and greater than twice that of the second most abundant nucleotide, it is assigned as the consensus nucleotide and given in upper case. If the second criterion is satisfied but not the first, then the nucleotide is shown in lower case (see Table 3). If the sum of the frequencies is greater than 75% (but neither satisfies the above), they are jointly assigned the status of the consensus.

C. Termination of Transcription and Polyadenylation

Filamentous fungal mRNAs tend to have heterogeneous 3' ends with several apparent polyadenylation sites, and this has been confirmed when cDNA clones have been sequenced (Sebald and Kruse, 1984; Kleene et al., 1987). The higher eukaryotic polyadenylation signal AATAAA (Proudfoot and Brownlee, 1976) is seen intact or degenerate preceding polyadenylation sites in some fungal genes (e.g., Clements and Roberts, 1986; Mullaney et al., 1985) but is more often absent. Consequently, the sequence requirements for transcription termination and polyadenylation of filamentous fungi are unknown.

D. Intervening Sequences

The majority of filamentous fungal genes sequenced thus far contain intervening sequences, or introns, and these range in size from 48 bp (Gwynne et al., 1987) to 398 bp (Sachs et al., 1989), though most are less than 100 bp. Although the positions of introns are usually conserved between equivalent or related genes of *Neurospora* and *Aspergillus*, there are instances of the presence of an intron in one species and the absence in the other (Sebald and Kruse, 1984; Ward and Turner, 1986; and see Gurr et al., 1987).

Sequences present at the 5' and 3' splice sites are highly conserved (Table 3) and are similar to those observed in yeast (Langford and Gallwitz, 1983) and in higher eukaryotes (Mount, 1982). In addition, a conserved sequence similar to the yeast conserved internal lariat sequence TACTAAC (Langford and Gallwitz, 1983) is present. The importance of the conserved splice sites and internal sequence can be inferred from the detrimental effect of mutation in these sequences in the *N. crassa cyt-21* and *cyc-1* genes (Kuiper et al., 1988; Stuart et al., 1987).

Although these sequences are similar to those seen in *S. cerevisiae* introns, this organism splices filamentous fungal introns very inefficiently (Innis et al., 1985; McKnight et al., 1985), and this is probably related to the stringent requirement for the TACTAAC sequence in yeast (Jacquier et al., 1986).

E. Codon Usage

All organisms tend to favor a subset of codons and this bias is generally most marked in highly expressed genes. Filamentous fungi have a bias in their codon usage (Ballance, 1986), though the degree of bias varies, it being particularly marked in *N. crassa* and less so in *A. nidulans*. The preferred codons are similar in all filamentous fungi for which genes have

Table 3 Frequency of Bases at Intron Splice Sites and the Internal Site of Lariat Formation

Organism	Base	5′ Splice site									Internal site of lariat formation									3′ Splice site						
N. crassa	G	35	62	0	0	7	3	61	5	5	15	14	40	1	0	27	0	1	7	8	3	5	0	0	63	27
	A	4	0	0	55	28	1	3	14		21	16	17	0	0	33	63	0	32	16	20	36	2	63	0	9
	T	10	0	61	1	12	1	45	20		15	22	5	61	2	1	0	4	12	13	22	5	24	0	0	7
	C	9	1	2	0	20	0	10	19		7	11	1	2	23	0	0	58	12	21	13	12	37	0	0	15
A. nidulans	G	34	67	0	0	11	5	59	7	4	13	10	26	3	2	23	0	2	21	9	9	11	1	0	67	24
	A	4	0	0	51	25	4	6	13		15	13	24	4	0	40	67	3	15	13	13	21	4	67	0	14
	T	13	0	67	3	21	4	44	17		15	31	13	6	63	3	0	11	22	17	17	10	31	0	0	12
	C	5	0	0	2	16	0	0	24		12	13	4	54	2	1	0	51	9	18	18	15	31	0	0	6
All filamentous fungi (n = 201)	G	97	200	0	0	35	16	182	18	26	46	38	97	4	7	73	0	7	40	25	20	32	2	0	201	71
	A	12	0	0	157	94	8	14	43		47	41	66	5	1	111	201	5	69	45	39	76	10	201	0	39
	T	36	0	198	3	6	46	10	141	52	54	90	26	20	188	9	0	29	51	49	69	27	83	0	0	28
	C	30	1	3	3	45	1	28	56		27	32	12	172	5	8	0	160	41	57	48	41	106	0	0	37
Consensus		G	G	T	A	a	G	T			t	$^{G}/_{A}$		C	T	A	A	C					$^{C}/_{T}$	A	G	

Data were compiled from the sequences listed in Appendix 1.

In some cases full information was not available (see Gurr et al., 1987).

The consensus sequences were assigned according to the criteria used in Table 2.

been sequenced. There is a preference for codons ending in pyrimidines, or where this is not possible, codons ending in G. This is also seen in *S. cerevisiae* (Bennetzen and Hall, 1982), though there are several cases where the codons preferred by *S. cerevisiae*, e.g., GAA for Glu and AGA for Arg, are different from those preferred by filamentous fungi; GAG and GGU or CGC, respectively (Ballance, 1986).

V. CONCLUSIONS

It is clear from the contents of this volume that recombinant DNA techniques will have an increasingly important impact on all aspects of industrial mycology over the next few years. There are now a wide variety of transformation selection systems available, and these should enable the application of these techniques to any fungus that can be grown in vitro. Generally, it is facile to transform an organism and produce stable transformants containing multiple copies of a gene and thereby influence the productivity of the fungus with respect to the specified protein or derivatives of it.

In addition to engineering an increase in gene copy number, it is possible to disrupt a gene by transformation, thereby creating a null mutation. This technique will be of use in the removal of unwanted contaminants from industrial fermentations and also in the manipulation of plant pathogenic fungi. A great deal of fundamental information on gene and genome structure will also become available from the utilization of recombinant DNA techniques in the genetic analysis of filamentous fungi.

ACKNOWLEDGMENTS

I would like to thank the Directors of Delta Biotechnology Limited for permission to publish this article and Helena Gordon, Joanne Middleton, and Sandra Wood for typing this manuscript.

APPENDIX 1

Sequences of the following genes were used in the compilation of data.

A. *Neurospora crassa*

actin	cited in Gurr et al. (1987)
ADP/ATP carrier	Arends and Sebald (1984)
am	Kinnaird and Fincham (1983)
ATP synthase delta subunit	Selbald and Kruse (1984)

ATP synthase proteolipid sub- unit	Viebrock et al. (1982)
con-8	Roberts and Yanofsky (1989)
cpc-1	Paluh et al. (1988)
crp-1	Kreader and Heckman (1987)
cya-4	Sachs et al. (1989)
cyc-1	Stuart et al. (1987)
cyclophilin	Tropschug et al. (1988)
cyt-18	Akins and Lambowitz (1987)
cyt-21	Kuiper et al. (1988)
H^+-ATPase	Hager et al. (1986)
his-3	Legerton and Yanofsky (1985)
histones H3,H4	Wouldt et al. (1983)
iron-sulphur subunit of ubiquinol-cytochrome C reductase	Harnish et al. (1985)
methallothionein	Munger et al. (1985)
porin	Kleene et al. (1987)
pyr-4	Newbury et al. (1986)
qa-1S	Huiet and Giles (1986)
qa-2,qa-3	Alton et al. (1982)
qa-4	Rutledge (1984)
qa-x, qa-y	cited in Gurr et al. (1987)
trp-1	Schechtman and Yanofsky (1983)
trp-3	Burns and Yanofsky (1989)
β-tubulin	Orbach et al. (1986)
vma-1	Bowman et al. (1988a)
vma-2	Bowman et al. (1988b)

B. *Aspergillus nidulans*

abaA	Mirabito et al. (1989)
actA	Fidel et al. (1988)
acuD	Ballance, Lewis, and Turner (un- published)
argB	Upshall et al. (1986)
adh3	McKnight et al. (1985)
alcA	Gwynne et al. (1987)
alcR	Felenbok et al. (1988)
aldA	Pickett et al. (1987)
amdS	Corrick et al. (1987)

aroM	Hawkins (1987)
benA	May et al. (1987)
bimG	Doonan and Morris (1989)
brlA	Adams et al. (1988)
gdhA	cited in Gurr et al. (1987)
gpdA	van Gorcom et al. (1987)
H2A	cited in Gurr et al. (1987)
oliC	Ward and Turner (1986)
pacA	MacRae et al. (1988)
pgkA	Clements and Roberts (1986)
prnB	Sophianopoulou and Scazzocchio (1989)
pykA	de Graaff and Visser (1988)
qutA	Beri et al. (1987)
qutB, qutD, qutG	Hawkins et al. (1988)
qutE	Da Silva et al. (1986)
tpiA	McKnight et al. (1986)
trpC	Mullaney et al. (1985)
tubA	cited in Gurr et al. (1987)
tubC	May et al. (1987)

C. Other Filamentous Fungi

1. *Aspergillus niger*

adhA	McKnight and Upshall (1987)
amyloglucosidase	Boel et al. (1984)
argB	Buxton et al. (1987)
gpdA	van Gorcom et al. (1987)
oliC	Ward et al. (1988)
pyrG	Wilson et al. (1988)
tpiA	McKnight and Upshall (1987)
trpC	Kos et al. (1988)

2. *Aspergillus oryzae*

TAKA-amylase	Boel et al. (1987)

3. *Colletotrichum capsici*

cutA	cited in Gurr et al. (1987)

4. *Colletotrichum gloesporioides*

cutA	cited in Gurr et al. (1987)

5. *Fusarium solani*
   cutA   Kolattukudy et al. (1985)

6. *Mucor meihei*
   Aspartyl protease   Gray et al. (1986)
   Lipase   Boel et al. (1987)

7. *Mucor pusillus*
   Aspartyl protease   Tonouchi et al. (1986)

8. *Mucor racemosus*
   TEF-1, TEF-2, TEF-3   Sundstorm et al. (1987)

9. *Penicillum chrysogenum*
   pgk1   van Solingen et al. (1988)
   pyrG   Cantoral et al. (1988)
   trpC   Penalva and Sanchez (1987)

10. *Penicillium patulum*
    FAS2   Weisner et al. (1988)

11. *Phanerochaete chrysosporium*
    Ligninase H   Tien and Tu (1987); Smith et al.
                                            (1988)
    CLG4, CLG5   de Boer et al. (1987)
    Ligninase 0282   Schalch et al. (1989)
    Ligninase V4   Schalch et al. (1989)

12. *Podospora anserina*
    ura5   Turcq and Begueret (1987)

13. *Rhizopus niveus*
    Aspartyl proteinase   Horiuchi et al. (1988)

14. *Rhizopus oryzae*
    Glucoamylase   Ashikari et al. (1986)

15. *Schizophyllum commune*
    1G2   Dons et al. (1984)

16. *Trichoderma reesei*
    CBHI   Shoemaker et al. (1983)
    CBHII   Chen et al. (1987)
    EGI   Penttila et al. (1986)
    egl3   Saloheimo et al. (1988)
    pgk   Vanhanen et al. (1989)

REFERENCES

Akins, R.A. and Lambowitz, A.M. (1985). General method for cloning *Neurospora crassa* nuclear genes by complementation of mutants. *Mol. Cell. Biol.* *5*: 2272-2278.

Akins, R.A. and Lambowitz, A.M. (1987). A protein required for splicing group I introns in *Neurospora* mitochondria is mitochondrial tyrosyl-tRNA synthetase or a derivative thereof. *Cell 50*: 331-345.

Alton, N.K., Buxton, F., Patel, V., Giles, N.H., and Vapnek, D. (1982). 5'-untranslated sequences of two structural genes in the *qa* cluster of *Neurospora crassa. Proc. Natl. Acad. Sci. USA 79*: 1955-1959.

Arends, H. and Sebald, W. (1984). Nucleotide sequence of the cloned mRNA and gene of the ADP/ATP carrier from *Neurospora crassa. EMBO J. 3*: 377-382.

Ashikari, T., Nakamura, N., Tanaka, Y., Kiuchi, N., Shibano, Y., Tanaka, T., Amachi, T., and Yoshizumi, H. (1986). *Rhizopus* raw-starch-degrading glucoamylase: its cloning and expression in yeast. *Agric. Biol. Chem. 50*: 957-964.

Ballance, D.J. (1986). Sequences important for gene expression in filamentous fungi. *Yeast 2*: 229-236.

Ballance, D.J., Buxton, F.P., and Turner, G. (1983). Transformation of *Aspergillus nidulans* by the orotidine-5'-phosphate decarboxylase gene of *Neurospora crassa. Biochem. Biophys. Res. Comm. 112*: 284-289.

Ballance, D.J. and Turner, G. (1985). Development of a high-frequency transforming vector for *Aspergillus nidulans. Gene 36*: 321-331.

Ballance, D.J. and Turner, G. (1986). Gene cloning in *Aspergillus nidulans*: isolation of the isocitrate lyase gene (*acuD*). *Mol. Gen. Genet. 202*: 271-275.

Barnes, D.E. and McDonald, D.W. (1986). Behaviour of recombinant plasmids in *Aspergillus nidulans*: structure and stability. *Curr. Genet. 10*: 767-775.

Baum, J.A., Geever, R., and Giles, N.H. (1987). Expression of qa-1F activator protein: identification of upstream binding sites in the *qa* gene cluster and localisation of the DNA-binding domain. *Mol. Cell. Biol. 7*:1256-1266.

Beggs, J.D. (1978). Transformation of yeast by a replicating hybrid plasmid. *Nature 275*: 104-109.

Begueret, J., Razanamparany, V., Perrot, M., and Barreau, C. (1984). Cloning gene *ura5* for the orotidylic acid pyrophosphorylase of the filamentous fungus *Podospora anserina*: transformation of protoplasts. *Gene 32*: 487-492.

Bennetzen, J.L. and Hall, B.D. (1982). Codon selection in yeast. *J. Biol. Chem. 257*: 3026-3031.

Benoist, C., O'Hare, K., Breathnach, R., and Chambon, P. (1980). The ovalbumin gene: sequence of putative control regions. *Nucleic. Acids Res. 8*:127-142.

Beri, R.K. and Turner, G. (1987). Transformation of *Penicillium chrysogenum* using the *Aspergillus nidulans amdS* gene as a dominant selective marker. *Curr. Genet. 11*: 639-641.

Beri, R.K., Whittington, H., Roberts, C.F., and Hawkins, A.R. (1987). Isolation and characterisation of the positively acting regulatory gene *QUTA* from *Aspergillus nidulans*. *Nucleic Acids Res. 15*: 7991-8001.

Berse, B., Dmochowska, A., Skrzypek, M., Weglenski, P., Bates, M.A., and Weiss, R.L. (1983). Cloning and characterisation of the ornithine carbamoyltransferase gene from *Aspergillus nidulans*. *Gene 25*: 109-117.

Binninger, D.M., Skrzynia, C., Pukkila, P.J., and Casselton, L.A. (1987). DNA-mediated transformation of the basidiomycete *Coprinus cinereus*. *EMBO J. 6*: 835-840.

Boeke, J.D., Lacroute, F., and Fink, G.R. (1984). A positive selection for mutants of yeast lacking orotidine-5′-phosphate decarboxylase activity in yeast: 5-fluoro-orotic acid resistance. *Mol. Gen. Genet. 197*: 345-346.

Boel, E., Hansen, M.T., Hjort, I., Hoegh, I., and Fiil, N.P. (1984). Two different types of intervening sequences in the glucoamylase gene from *Aspergillus niger*. *EMBO J. 3*: 1581-1585.

Boel, E., Christensen, T., and Woldike, H.F. (1987). Process for the production of protein products in *Aspergillus*. European Patent Application #0 238 023 A2.

de Boer, H.A., Zhang, Y.Z., Collins, C., and Reddy, C.A. (1987). Analysis of nucleotide sequences of two ligninase cDNA's from a white-rot filamentous fungus, *Phanerochaete chrysosporium*. *Gene 60*: 93-102.

Bowman, E.J., Tenney, K., and Bowman, B.J. (1988a). Isolation of genes encoding the *Neurospora* vacuolar ATPase. Analysis of *vma-1* encoding the 67-kDa subunit reveals homology to other ATPases. *J. Biol. Chem. 263*: 13994-14001.

Bowman, B.J., Allen, R., Wechser, M.A., and Bowman, E.J. (1988b). Isolation of genes encoding the *Neurospora* vacuolar ATPase. Analysis of *vma-2* encoding the 57-kDa polypeptide and comparison to *vma-1*. *J. Biol. Chem. 263*: 14002-14007.

Burns, D.M. and Yanofsky, C. (1989). Nucleotide sequence of the *Neurospora crassa trp3* gene encoding tryptophan synthetase and comparison of the *trp3* polypeptide with its homologs in *Saccharomyces cerevisiae* and *Escherichia coli*. *J. Biol. Chem. 264*: 3840-3848.

Buxton, F.P., Gwynne, D.I., Garven, S., Sibley, S., and Davies, R.W. (1987). Cloning and molecular analysis of the ornithine carbamoyl transferase gene of *Aspergillus niger*. *Gene 60*: 255-266.

Cantoral, J.M., Barredo, J.L., Alvarez, E., Diez, B., and Martin, J.F. (1988). Nucleotide sequence of the *Penicillium chrysogenum pyrG* (orotidine-5′-phosphate decarboxylase) gene. *Nucleic Acids Res. 16*: 8177.

Case, M.E. (1986). Genetical and molecular analysis of *qa-2* transformants in *Neurospora crassa*. *Genetics 113*: 569-587.

Case, M.E., Schweizer, M., Kushner, S.R., and Giles, N.H. (1979). Efficient transformation of *Neurospora crassa* by utilising hybrid plasmid DNA. *Proc. Natl. Acad. Sci. USA 76*: 5259-5363.

Charles, I.G., Keyte, J.W., Brammer, W.J., Smith, M., and Hawkins, A.R. (1986). The isolation and nucleotide sequence of the complex AROM locus of *Aspergillus nidulans*. *Nucleic Acids Res. 14*: 2201-2213.

Chen, C.M., Gritzali, M., and Stafford, D.W. (1987). Nucleotide sequence and deduced primary structure of cellobiohydrolase II from *Trichoderma reesei. Bio/Technology 5*: 274-278.

Cigan, A.M. and Donahue, T.F. (1987). Sequence and structural features associated with translational initiator regions in yeast—a review. *Gene 59*: 1-18.

Clark, L. and Carbon, J. (1976). A colony bank containing synthetic ColE1 hybrid plasmids representative of the entire *E. coli* genome. *Cell 9*: 91-99.

Clements, J.M. and Roberts, C.F. (1986). Transcription and processing signals in the 3-phosphoglycerate kinase (PGK) gene from *Aspergillus nidulans. Gene 44*: 97-105.

Cooley, R.N., Shaw, R.K., Franklin, F.C.H., and Caten, C.E. (1988). Transformation of the phytopathogenic fungus *Septoria nodorum* to hygromycin β resistance. *Curr. Genet. 13*: 383-389.

Corrick, C.M., Twomey, A.P., and Hynes, M.J. (1987). The nucleotide sequence of the *amdS* gene of *Aspergillus nidulans* and the molecular characterisation of 5' mutations. *Gene 53*: 63-71.

Cullen, D. and Leong, S. (1986). Recent advances in the molecular genetics of industrial filamentous fungi. *Trends Biotechnol. 4*: 285-288.

Daboussi, M.J., Djeballi, A., Gerlinger, C., Blaiseau, P.L., Bouvier, I., Cassan, M., Lebrun, M.H., Parisot, D., and Brygoo, Y. (1989). Transformation of seven species of filamentous fungi using the nitrate reductase gene of *Aspergillus nidulans. Curr. Genet. 15*: 453-456.

Da Silva, A.J.F., Whittington, H., Clements, S., Roberts, C., and Hawkins, A.R. (1986). Sequence analysis and transformation by the catabolic 3-dehydroquinase (*QUTE*) gene from *Aspergillus nidulans. Biochem. J. 240*: 481-488.

Dhawale, S.S. and Marzluf, G.A. (1985). Transformation of *Neurospora crassa* with circular and linear DNA and analysis of the fate of the transforming DNA. *Curr. Genet. 10*: 205-212.

Dhawale, S.S., Paietta, J.V., and Marzluf, G.A. (1984). A new, rapid and efficient transformation procedure for *Neurospora. Curr. Genet. 8*: 77-79.

Dons, J.J.M., Mulder, G.H., Rouwendal, G.J.A., Springer, J.A., Bremer, W., and Wessels, J.G.H. (1984). Sequence analysis of a split gene involved in fruiting from the fungus *Schizophyllum commune. EMBO J. 3*: 2101-2106.

Doonan, J.H. and Morris, N.R. (1989). The *bimG* gene of *Aspergillus nidulans*, required for completion of anaphase, encodes a homolog of mammalian phosphoprotein phosphatase 1. *Cell 57*: 987-996.

Felenbok, B., Sequeval, D., Mathieu, M., Sibley, S., Gwynne, D.I., and Davies, R.W. (1988). The ethanol regulon in *Aspergillus nidulans*: characterisation and sequence of the positive regulatory gene *alcR. Gene 73*: 385-396.

Fu, Y.H., Paietta, J.V., Mannix, D.G., and Marzluf, G.A. (1989). *cys-3*, the positive-acting sulphur regulatory gene of *Neurospora crassa*, encodes a protein with a putative leucine zipper DNA-binding element. *Mol. Cell. Biol. 9*: 1120-1127.

Gannon, F., O'Hara, K., Perrin, F., Le Pennec, J.P., Benoist, C., Cochet, M., Breathnach, R., Royal, A., Garaspin, A., Cami, B., and Chambon, P. (1979). Organisation and sequences at the 5' end of a cloned complete ovalbumin gene. *Nature 278*: 428-434.

Geisen, R. and Leistner, L. (1989). Transformation of *Penicillum nalgiovense* with the *amdS* gene of *Aspergillus nidulans. Curr. Genet. 15*: 307-309.

van Gorcom, R.F.M., Punt, P.J., de Ruiter-Jacobs, Y.M.R., Dingemanse, M.A., Kuijvenhoven, A., Pouwels, P.H., and van den Hondel, C.A.M.J.J. (1987). Isolation and characterisation of the highly expressed GPD genes of *Aspergillus nidulans* and *Aspergillus niger*. Poster presentation at the 19th Lunteren Lecture on Molecular Biology: Molecular Genetics of Yeast and Filamentous Fungi and its Impact on Biotechnology.

de Graaff, L. and Visser, J. (1988). Structure of the *Aspergillus nidulans* pyruvate kinase gene. *Curr. Genet. 14*: 553-560.

Grant, D.M., Lambowitz, A.M., Rambosek, J.A., and Kinsey, J.A. (1984). Transformation of *Neurospora crassa* with recombinant plasmids containing the cloned glutamate dehydrogenase (*am*) gene: evidence for autonomous replication of the transforming plasmid. *Mol. Cell. Biol. 4*: 2041-2051.

Gray, G.L., Hayenga, K., Cullen, D., Wilson, L.J., and Norton, S. (1986). Primary structure of *Mucor miehei* aspartyl protease: evidence for a zymogen intermediate. *Gene. 48*: 41-53.

Gurr, S.J., Unkles, S.E., and Kinghorn, J.R. (1987). The structure and organisation of nuclear genes of filamentous fungi. In *Gene Structure in Eukaryotic Microbes*, Vol. 22, Kinghorn, J.R., ed. Special publication of the Society for General Microbiology. IRL Press, Oxford, p. 93-139.

Gwynne, D.I., Buxton, F.P., Sibley, S., Davies, R.W., Lockington, R.A., Scazzocchio, C., and Sealy-Lewis, H.M. (1987). Comparison of the *cis*-acting control regions of two coordinately controlled genes involved in ethanol utilisation in *Aspergillus nidulans. Gene 51*: 205-216.

Hager, K.M., Mandala, S.M., Davenport, S.W., Speicher, D.W., Benz, E.J., and Slayman, C.W. (1986). Amino acid sequence of the plasma membrane ATPase of *Neurospora crassa*: deduction from genomic and cDNA sequences. *Proc. Natl. Acad. Sci. USA 83*: 7693-7697.

Hamer, J.E. and Timberlake, W.E. (1987). Functional organisation of the *Aspergillus nidulans trpC* promoter. *Mol. Cell. Biol. 7*: 2352-2359.

Harnish, U., Weiss, H., and Sebald, W. (1985). The primary structure of the iron-sulfur subunit of ubiquinol-cytochrome *c* reductase from *Neurospora*, determined by cDNA and gene sequencing. *Eur. J. Biochem. 149*: 95-99.

van Hartingsveldt, W., Mattern, I.E., van Zeijl, C.M.J., Pouwels, P.H., and van den Hondel, C.A.M.J.J. (1987). Development of a homologous transformation system for *Aspergillus niger* based on the *pyr4* gene. *Mol. Gen. Genet. 206*: 71-75.

Hawkins, A.R. (1987). The complex *Arom* locus of *Aspergillus nidulans*. *Curr. Genet. 11*: 491-498.

van Heeswijck, R. (1986). Autonomous replication of plasmids in *Mucor* transformants. *Carlsberg Res. Commun. 51*: 433-443.

van Heeswijck, R. and Roncero, M.I.G. (1984). High frequency transformation of *Mucor* with recombinant plasmid DNA. *Carlsberg Res. Commun. 49*: 691-702.

Henson, J.M., Blake, N.K., and Pilgeram, A.L. (1988). Transformation of *Gaeumannomyces graminis* to benomyl resistance. *Curr. Genet. 14*: 113-117.

Hinnen, A., Hicks, J.B., and Fink, G.R. (1978). Transformation of yeast. *Proc. Natl. Acad. Sci. USA 75*: 1929-1933.

Holden, D.W., Wang, J., and Leong, S.A. (1988). DNA-mediated transformation of *Ustilago hordei* and *Ustilago nigra*. *Physiol. Mol. Pl. Pathol. 33*: 235-239.

Horiuchi, H., Yanai, K., Okazaki, T., Takagi, M., and Yano, K. (1988). Isolation and sequencing of a genomic clone encoding aspartic proteinase of *Rhizopus niveus*. *J. Bacteriol. 170*: 272-278.

Huang, D., Bhaire, S., and Staples, R.C. (1989). A transformation procedure for *Botryotinia squamosa*. *Curr. Genet. 15*: 411-414.

Huiet, L. and Giles, N.H. (1986). The *qa* repressor gene of *Neurospora crassa*: wild-type and mutant nucleotide sequences. *Proc. Natl. Acad. Sci. USA 83*: 3381-3385.

Hynes, M.J., Corrick, C.M., and King, J.A. (1983). Isolation of genomic clones containing the *amdS* gene of *Aspergillus nidulans* and their use in the analysis of structural and regulatory mutations. *Mol. Cell. Biol. 3*: 1430-1439.

Innis, M.A., Holland, M.J., McCabe, P.C., Cole, G.F., Wittman, V.P., Tal, R., Watt, K.W.K., Gelfand, D.H., Holland, J.P., and Meade, J.H. (1985). Expression, glycosylation and secretion of an *Aspergillus* glucoamylase by *Saccharomyces cerevisiae*. *Science 228*: 21-26.

Jacquier, A., Rodriguez, J.R., and Rosbash, M. (1986). A quantitative analysis of 5' junction and TACTAAC box mutants and mutant combinations on yeast mRNA splicing. *Cell 43*: 423-430.

Johnstone, I.L., Hughes, S.G., and Clutterbuck, A.J. (1985). Cloning an *Aspergillus nidulans* developmental gene by transformation. *EMBO J. 4*: 1307-1311.

Keesey, J.K. and DeMoss, J.A. (1982). Cloning of the *trp-1* gene from *Neurospora crassa* by complementation of a *trpC* mutation in *Escherichia coli*. *J. Bacteriol. 152*: 954-958.

Kelly, J.M. and Hynes, M.J. (1985). Transformation of *Aspergillus niger* by the *amdS* gene of *Aspergillus nidulans*. *EMBO J. 4*: 475-479.

Kelly, J.M. and Hynes, M.J. (1987). Multiple copies of the *amdS* gene of *Aspergillus nidulans* cause titration of *trans*-acting regulatory proteins. *Curr. Genet. 12*: 21-31.

Kim, S.Y. and Marzluf, G.A. (1988). Transformation of *Neurospora crassa* with the *trp-1* gene and the effect of host strain upon the fate of the transforming DNA. *Curr. Genet. 13*: 65-70.

Kinnaird, J.H. and Fincham, J.R.S. (1983). The complete nucleotide sequence of the *Neurospora crassa am* (NADP-specific glutamate dehydrogenase) gene. *Gene 26*: 253-260.

Kinnaird, J.H., Keighren, M.A., Kinsey, J.A., Eaton, M., and Fincham, J.R.S. (1982). Cloning of the *am* (glutamate dehydrogenase) gene of *Neurospora crassa* through the use of a synthetic DNA probe. *Gene 20*: 387-396.

Kistler, H.C. and Benny, U.K. (1988). Genetic transformation of the fungal plant wilt pathogen, *Fusarium oxysporum. Curr. Genet. 13*: 145-149.

Kleene, R., Pfanner, N., Pfaller, R., Link, T.A., Sebald, W., Neupert, W., and Tropschug, M. (1987). Mitochondrial porin of *Neurospora crassa*: cDNA cloning, in vitro expression and import into mitochondria. *EMBO J. 6*: 2627-2633.

Kolattukudy, P.E., Soliday, C.L., Woloshuk, C.P., and Crawford, M. (1985). Molecular biology of the early events in the fungal penetration into plants. In *Molecular Genetics of Filamentous Fungi*, UCLA Symposia on Molecular and Cellular Biology, New Series, Vol. 34, Timberlake, W.E., ed., Alan R. Liss, New York, p. 421-438.

Kos, T., Kuijvenhoven, A., Hessing, H.G.M., Pouwels, P.H., and van den Hondel, C.A.M.J.J. (1988). Nucleotide sequence of the *Aspergillus niger trpC* gene: structural relationship with analogous genes of other organisms. *Curr. Genet. 13*: 137-144.

Kozak, M. (1978). How do eukaryotic ribosomes select initiation regions in messenger RNA? *Cell 15*: 1109-1123.

Kozak, M. (1983). Comparison of initiation of protein synthesis in procaryotes, eucaryotes, and organelles. *Microbiol. Rev. 47*: 1-45.

Kozak, M. (1987). An analysis of 5'-noncoding sequences from 669 vertebrate messenger RNAs. *Nucleic Acids Res. 15*: 8125-8148.

Kreader, C.A. and Heckman, J.E. (1987). Isolation and characterisation of a *Neurospora crassa* ribosomal protein gene homologous to *CYH2* of yeast. *Nucleic Acids Res. 15*: 9027-9042.

Krumlauf, R. and Marzluf, G.A. (1980). Genome organisation and characterisation of the repetitive and inverted repeat DNA sequences in *Neurospora crassa. J. Biol. Chem. 225*: 1138-1145.

Kuiper, M.T.R., Akins, R.A., Holtrop, M., de Vries, H., and Lambowitz, A.M. (1988). Isolation and analysis of the *Neurospora crassa Cyt-21* gene. *J. Biol. Chem. 263*: 2840-2847.

Langford, C.J. and Gallwitz, D. (1983). Evidence for an intron-contained sequence required for the splicing of yeast RNA polymerase II transcripts. *Cell 33*: 519-527.

Legerton, T.L. and Yanofsky, C. (1985). Cloning and characterisation of the multifunctional *his-3* gene of *Neurospora crassa. Gene 39*: 129-140.

Lockington, R.A., Sealy-Lewis, H.M., Scazzocchio, C., and Davies, R.W. (1985). Cloning and characterisation of the ethanol utilisation regulon in *Aspergillus nidulans. Gene 33*: 137-149.

MacRae, W.D., Buxton, F.P., Sibley, S., Garven, S., Gwynne, D.I., Davies, R.W., and Arst, H.N. Jr. (1988). A phosphate-repressible acid phosphatase gene from *Aspergillus niger*: its cloning, sequencing and transcriptional analysis. *Gene 71*: 339-348.

Manavathu, E.K., Suryanarayana, K., Hasnain, S.E., Leung, M., Lau, Y.F., and Leung, W.C. (1987). Expression of *Herpes simplex* virus thymidine kinase gene in aquatic filamentous fungus *Achlya ambisexualis*. *Gene 57*: 53-59.

Maniatis, T., Fritsch, E.F., and Sambrook, J. (1982). *Molecular Cloning: A Laboratory Manual*. Cold Spring Harbor Laboratory, Cold Spring Harbor, NY.

Marek, E.T., Scharal, C.L., and Smith, D.A. (1989). Molecular transformation of *Fusarium solani* with an antibiotic resistance marker having no fungal DNA homology. *Curr. Genet. 15*: 421-428.

Mattern, I.E., Unkles, S., Kinghorn, J.R., Pouwels, P.H., and van de Hondel, C.A.M.J.J. (1987). Transformation of *Aspergillus oryzae* using the *A. niger pyrG* gene. *Mol. Gen. Genet. 210*: 460-461.

May, G.S., Gambino, J., Weatherbee, J.A., and Morris, N.R. (1985). Identification and functional analysis of beta-tubulin genes by site specific integrative transformation in *Aspergillus nidulans*. *J. Cell. Biol. 101*: 712-719.

May, C.S., Tsang, M.L.S., Smith, H., Fidel, S., and Morris, N.R. (1987). *Aspergillus nidulans* β-tubulin genes are unusually divergent. *Gene 55*: 231-243.

McKnight, G.L., Kato, H., Upshall, A., Parker, M.D., Saari, G., and O'Hara, P.J. (1985). Identification and molecular analysis of a third *Aspergillus nidulans* alcohol dehydrogenase. *EMBO J. 4*: 2093-2099.

McKnight, G.L., O'Hara, P.J., and Parker, M.L. (1986). Nucleotide sequence of the triosephosphate isomerase gene from *Aspergillus nidulans*: implications for a differential loss of introns. *Cell 46*: 143-147.

McKnight, G.L. and Upshall, A. (1987). Expression of higher eucaryotic genes in *Aspergillus*. International Patent Application #WO 87/04464.

Miller, B.L., Miller, K.Y., and Timberlake, W.E. (1985). Direct and indirect gene replacements in *Aspergillus nidulans*. *Mol. Cell. Biol. 5*: 1714-1721.

Miller, B.L., Miller, K.Y., Roberti, K.A., and Timberlake, W.E. (1987). Position-dependent and -independent mechanisms regulate cell-specific expression of the SpoC1 gene cluster of *Aspergillus nidulans*. *Mol. Cell. Biol. 7*: 427-434.

Mirabito, P.M., Adams, T.H., and Timberlake, W.E. (1989). Interactions of three sequentially expressed genes control temporal and spatial specificity in *Aspergillus* development. *Cell 57*: 859-868.

Mount, S.M. (1982). A catalogue of splice junction sequences. *Nucleic Acids Res. 10*: 459-472.

Mullaney, E.J., Hamer, J.E., Roberti, K.A., Yelton, M.M., and Timberlake, W.E. (1985). Primary structure of the *trpC* gene from *Aspergillus nidulans*. *Mol. Gen. Genet. 199*: 37-45.

Munger, K., Germann, U.A., and Lerch, K. (1985). Isolation and structural organisation of the *Neurospora crassa* copper metallothionein gene. *EMBO J. 4*: 2665-2668.

Munoz-Rivas, A., Specht, C.A., Drummond, B.J., Froeliger, E., Novotny, C.P., and Ullrich, R.C. (1986). Transformation of the basidiomycete *Schizophyllum commune. Mol. Gen. Genet. 205*: 103-106.

Newbury, S.F., Glazebrook, J.A., and Radford, A. (1986). Sequence analysis of the *pyr-4* (orotidine 5'-P decarboxylase) gene of *Neurospora crassa. Gene 43*: 51-58.

Oliver, R.P., Roberts, I.N., Harling, R., Kenyon, L., Punt, P.J., Dingemanse, M.A., and van den Hondel, C.A.M.J.J. (1987). Transformation of *Fulvia fulva*, a fungal pathogen of tomato, to hygromycin B resistance. *Curr. Genet. 12*: 231-233.

Orbach, M.J., Porro, E.B., and Yanofsky, C. (1986). Cloning and characterisation of the gene for β-tubulin from a benomyl-resistant mutant of *Neurospora crassa* and its use as a dominant selectable marker. *Mol. Cell. Biol. 6*: 2452-2461.

Osmani, S.A., Engle, D.B., Doonan, J.H., and Morris, N.R. (1988). Spindle formation and chromatin condensation in cells blocked at interphase by mutation of a negative cell cycle control gene. *Cell 52*: 241-251.

Paluh, J.L., Orbach, M.S., Legerton, T.L., and Yanofsky, C. (1988). The cross pathway control gene of *Neurospora crassa, cpc-1,* encodes a protein similar to GCN4 of yeast and the DNA binding domain of the oncogene v-*jun*-encoded protein. *Proc. Natl. Acad. Sci. USA 85*: 3728-3732.

Parsons, K.A., Chumley, F.G., and Valent, B. (1987). Genetic transformation of the fungal pathogen responsible for rice blast disease. *Proc. Natl. Acad. Sci. USA 84*: 4161-4165.

Penalva, M.A. and Sanchez, F. (1987). The complete nucleotide sequence of the *trpC* gene from *Penicillium chrysogenum. Nucleic Acids Res. 15*: 1874.

Penalva, M.A., Tourino, A., Patino, C., Sanchez, F., Fernandez, Sousa, J.M., and Rubio, V. (1985). Studies on transformation of *Cephalosporium acremonium*. In *Molecular Genetics of Filamentous Fungi*, UCLA Symposia on Molecular and Cellular Biology, New Series, Vol. 34, Timberlake, W.E., ed., Alan R. Liss, New York, p. 59-68.

Penttila, M.E., Nevalainen, K.M.H., Raynal, A., and Knowles, J.K.C. (1984). Cloning of *Aspergillus niger* genes in yeast. Expression of the gene coding *Aspergillus* β-glucosidase. *Mol. Gen. Genet. 194*: 494-499.

Penttila, M., Lehtovaara, P., Nevalhainen, H., Bhikhabhai, R., and Knowles, J. (1986). Homology between cellulase genes of *Trichoderma reesei*: complete nucleotide sequence of the endoglucanase I gene. *Gene 45*: 253-263.

Penttila, M., Nevalainen, H., Ratto, M., Salminen, E., and Knowles, J. (1987). A versatile transformation system for the cellulolytic filamentous fungus *Trichoderma reesei. Gene 61*: 155-164.

Perrot, M., Barreau, C., and Begueret, J. (1987). Nonintegrative transformation in the filamentous fungus *Podospora anserina*: stabilisation of a linear vector by the chromosomal ends of *Tetrahymena thermophila. Mol. Cell. Biol. 7*: 1725-1730.

Pickett, M., Gwynne, D.I., Buxton, F.P., Elliot, R., Davies, R.W., Lockington, R.A., Scazzocchio, C., and Sealy-Lewis, H.M. (1987). Cloning and characterisation of the *aldA* gene of *Aspergillus nidulans*. *Gene 51*: 217-226.

Proudfoot, N.J. and Brownless, G.G. (1976). 3'-Noncoding region sequences in eukaryotic messenger RNA. *Nature 263*: 211-214.

Punt, P.J., Oliver, R.P., Dingemanse, M.A., Pouwels, P.H., and van den Hondel, C.A.M.J.J. (1987). Transformation of *Aspergillus* based on the hygromycin B resistance marker from *Escherichia coli*. *Gene 56*: 117-124.

Roberts, A.N., Berlin, V., Hager, K.M., and Yanofsky, C. (1988). Molecular analysis of a *Neurospora crassa* gene expressed during conidation. *Mol. Cell. Biol. 8*: 2411-2418.

Roberts, A.N. and Yanofsky, C. (1989). Genes expressed during conidation in *Neurospora crassa*: characterisation of *con-8*. *Nucleic Acids Res. 17*: 197-214.

Rothstein, R.J. (1983). One step gene disruption in yeast. *Methods Enzymol. 101*: 202-211.

Rutledge, B.J. (1984). Molecular characterisation of the *qa-4* gene of *Neurospora crassa*. *Gene 32*: 275-287.

Sachs, M.S., Bertrand, H., Metzenberg, R.L., and RajBhandary, U.L. (1989). Cytochrome oxidase subunit V gene of *Neurospora crassa*: DNA sequences, chromosomal mapping, and evidence that the *cya-4* locus specifies the structural gene for subunit V. *Mol. Cell. Biol. 9*: 566-577.

Saloheimo, M., Lehtovaara, P., Penttila, M., Teeri, T.T., Stahlberg, J., Johansson, G., Petterson, G., Claeyssens, M., Tomme, P., and Knowles, J.K.C. (1988). EGIII, a new endoglucanase from *Trichoderma reesei*: the characterization of both gene and enzyme. *Gene 63*: 11-21.

Schalch, H., Gaskell, J., Smith, T.L., and Cullen, D. (1989). Molecular cloning and sequences of lignin peroxidase genes of *Phanerochaete chrysosporium*. *Mol. Cell. Biol. 9*: 2743-2747.

Schechtman, M.G. and Yanofsky, C. (1983). Structure of the trifunctional *trp-1* gene from *Neurospora crassa* and its aberrant expression in *Escherichia coli*. *J. Mol. Appl. Genet. 2*: 83-99.

Schweizer, M., Case, M.E., Dykstra, C.C., Giles, N.H., and Kushner, S.R. (1981). Cloning the quinic acid (*qa*) cluster from *Neurospora crassa*: identification of recombinant plasmids containing both *qa-2+* and *qa-3+*. *Gene 14*: 23-32.

Sebald, W. and Kruse, B. (1984). Nucleotide sequence of the nuclear genes for the proteolipid and delta subunit of the mitochondrial ATP synthase from *Neurospora crassa*. In *H^+-ATPase (ATP synthase): Structure, Function, Biogenesis. The F_0F_1 Complex of Coupling Membranes*, Papa, S., Altendorf, L., Ernster, L., and Packer, L., eds., Adriatica Editrice, Bari, p. 67-75.

Selker, E.U. Cambereri, E.B., Jensen, B.C., and Haack, K.R. (1987). Rearrangement of duplicated DNA in specialised cells of *Neurospora*. *Cell 51*: 741-752.

Shoemaker, S., Schweikart, V., Ladner, M., Gelfand, D., Kwok, S., Myambok, K., and Innis, M. (1983). Molecular cloning of exocellobiohydrase I derived from *Trichoderma reesei* strain L27. *Bio/Technology 1*: 691-696.

Smith, T.L., Schalch, H., Gaskell, J., Convert, S., and Cullen, D. (1988). Nucleotide sequence of a ligninase gene from *Phanerochaete chrysosporium*. *Nucleic Acids Res. 16*: 1219.

van Solingen, P., Muurling, H., Koekman, B., and van den Berg, J. (1988). Sequence of the *Penicillium chrysogenum* phosphoglycerate kinase gene. *Nucleic Acids Res. 16*: 11823.

Sophianopoulou, V. and Scazzocchio, C. (1989). The proline transport protein of *Aspergillus nidulans* is very similar to amino acid transporters of *Saccharomyces cerevisiae*. *Mol. Microbiol. 3*: 705-714.

Stahl, U., Leitner, E., and Esser, K. (1987). Transformation of *Penicillium chrysogenum* by a vector containing a mitochondrial origin of replication. *Appl. Microbiol. Biotechnol. 26*: 237-241.

Struhl, K. (1983). The new yeast genetics. *Nature 305*: 391-397.

Stuart, R.A., Neupert, W., and Tropschug, M. (1987). Deficiency in mRNA splicing in a cytochrome *c* mutant of *Neurospora crassa*: importance of carboxy terminus for import of apocytochrome *c* into mitochondria. *EMBO J. 6*: 2131-2137.

Sundstorm, P., Lira, L.M., Linz, J.E., and Sypherd, P.S. (1987). Sequence analysis of the EF-1α gene family of *Mucor racemosus*. *Nucleic Acids Res. 15*: 9997-10006.

Tien, M. and Tu, C.D. (1987). Cloning and sequencing of a cDNA for ligninase from *Phanerochaete chrysosporium*. *Nature 326*: 520-524.

Tilburn, J., Scazzocchio, C., Taylor, G.G., Zabicky-Zissman, J.H., Lockington, R.A., and Davies, R.W. (1983). Transformation by integration in *Aspergillus nidulans*. *Gene 26*: 295-321.

Timberlake, W.E. (1978). Low repetitive DNA content in *Aspergillus nidulans*. *Science 202*: 973-975.

Thomas, M.D. and Kenerley, C.M. (1989). Transformation of the mycoparasite *Gliocladium*. *Curr. Genet. 15*: 415-420.

Tonouchi, N., Shown, H., Uozumi, T., and Beppu, T. (1986). Cloning and sequencing of a gene for *Mucor* rennin, an aspartate protease from *Mucor pusillus*. *Nucleic Acids Res. 14*: 7557-7568.

Tropschug, M., Nicholson, D.W., Hartl, F.U., Kohler, H., Pfanner, N., Wachter, E., and Neupert, W. (1988). Cyclosporin A-binding protein (cyclophilin) of *Neurospora crassa*. *J. Biol. Chem. 263*: 14433-14440.

Turcq, B. and Begueret, J. (1987). The *ura5* gene of the filamentous fungus *Podospora anserina*: nucleotide sequence and expression in transformed strains. *Gene 53*: 201-209.

Turgeon, B.G., Garber, R.C., and Yoder, O.C. (1985). Transformation of the fungal maize pathogen *Cochliobolus heterostrophus* using the *Aspergillus nidulans amdS* gene. *Mol. Gen. Genet. 201*: 450-453.

Turgeon, B.G., Garber, R.C., and Yoder, O.C. (1987). Development of a fungal transformation system based on selection of sequences with promoter activity. *Mol. Cell. Biol. 7*: 3297-3305.

Upshall, A. (1986a). Filamentous fungi in biotechnology. *BioTechniques 4*: 158-166.

Upshall, A. (1986b). Genetic and molecular characterisation of *arg*B⁺ transformants of *Aspergillus nidulans. Curr. Genet. 10*: 593-599.

Upshall, A., Gilbert, T., Saari, G., O'Hara, P.J., Berse, B., Weglenski, P., Miller, K., and Timberlake, W.E. (1986). Molecular analysis of the *argB* gene of *Aspergillus nidulans. Mol. Gen. Genet. 204*: 349-351.

Vanhanen, S., Penttila, M., Lehtovaara, P., and Knowles, J. (1989). Isolation and characterisation of the 3-phosphoglycerate kinase gene (*pgk*) from the filamentous fungus *Trichoderma reesei. Curr. Genet. 15*: 181-186.

Viebrock, A., Perz, A., and Sebald, W. (1982). The imported preprotein of the proteolipid subunit of the mitochondrial ATP synthase from *Neurospora crassa.* Molecular cloning and sequencing of the mRNA. *EMBO J. 1*: 565-571.

Vollmer, S.J. and Yanofsky, C. (1986). Efficient cloning of genes of *N. crassa. Proc. Natl. Acad. Sci. USA 83*: 4869-4873.

Wang, J., Holden, D.W., and Leong, S.A. (1988). Gene transfer system for the phytopathogenic fungus *Ustilago maydis. Proc. Natl. Acad. Sci. USA 85*: 865-869.

Ward, M. and Turner, G. (1986). The ATP synthase subunit gene of *Aspergillus nidulans*: sequence and transcription. *Mol. Gen. Genet. 205*: 331-338.

Ward, M., Wilkinson, B., and Turner, G. (1986). Transformation of *Aspergillus nidulans* with a cloned oligomycin-resistant ATP synthase subunit 9 gene. *Mol. Gen. Genet. 202*: 265-270.

Ward, M., Wilson, L.J., Carmona, C.L., and Turner, G. (1988). The *oliC3* gene of *Aspergillus niger*: isolation, sequence and use as a selectable marker for transformation. *Curr. Genet. 14*: 37-42.

Wernars, K., Goosen, T., Swart, K., and van den Broek, H.W.S. (1986). Genetic analysis of *Aspergillus nidulans amdS*⁺ transformants. *Mol. Gen. Genet. 205*: 312-317.

Wiesner, P., Beck, J., Beck, K.-F., Ripka, S., Muller, G., Lucke, S., and Schweizer, E. (1988). Isolation and sequence analysis of the fatty acid synthetase *FAS2* gene from *Penicillium patulum. Eur. J. Biochem. 177*: 69-79.

Wilson, L.J., Carmona, C.L., and Ward, M. (1988). Sequence of the *Aspergillus niger pyrG* gene. *Nucleic Acids Res. 16*: 2339.

Wöstemeyer, A., Burmester, A., and Weigel, C. (1987). Neomycin resistance as a dominantly selectable marker for transformation of the zygomycete *Absidia glauca. Curr. Genet. 12*: 625-627.

Woudt, L.P., Pastink, A., Kempers-Veenstra, A.E., Jansen, A.E.M., Mager, W.H., and Planta, R.J. (1983). The genes coding for histones, H3 and H4 in *Neurospora crassa* are unique and contain intervening sequences. *Nucleic Acids Res. 11*: 5347-5360.

Yelton, M.M., Hamer, J.E., de Souza, E.R., Mullaney, E.J., and Timberlake, W.E. (1983). Development and regulation of the *Aspergillus nidulans trpC* gene. *Proc. Natl. Acad. Sci. USA 81*: 1470-1474.

Yelton, M.M., Hamer, J.E., and Timberlake, W.E. (1984). Transformation of *Aspergillus nidulans* by using a *trpC* plasmid. *Proc. Natl. Acad. Sci. USA 81*: 1470-1474.

Yelton, M.M., Timberlake, W.E., and van den Hondel, C.A.M.J.J. (1985). A cosmid for selecting genes by complementation in *Aspergillus nidulans*: selection of the developmentally regulated *yA* locus. *Proc. Natl. Acad. Sci. USA 82*: 834-838.

Zimmerman, C.R., Orr, W.C., Leclerc, R.F., Bernard, E.C., and Timberlake, W.E. (1980). Molecular cloning and selection of genes regulated in *Aspergillus* development. *Cell 21*: 709-715.

2

Molecular Manipulation of and Heterologous Protein Secretion from Filamentous Fungi

Alan Upshall, Ashok A. Kumar, Kenneth Kaushansky, and Gary L. McKnight
ZymoGenetics, Inc., Seattle, Washington

I. INTRODUCTION

Since the development of DNA-mediated transformation of *Aspergillus nidulans* (Ballance et al., 1983) and *Neurospora crassa* (Case et al., 1979), molecular methods have been successfully applied to the analysis and exploitation of many filamentous fungal species. Transformation was the critical step, since without the ability to introduce DNA into genomes little could be achieved, either academically or commercially, by the routine cloning and manipulation of individual genes. The diversification into many filamentous fungal species is because of their fundamental importance in the agricultural, chemical, and pharmaceutical industries. Plant pathogens are a considerable source of annual revenue loss and have unique experimental problems concerning the requirement for an interaction with a host. Many species do not grow in culture and molecular methods offer a way to investigate their biology (e.g., Upshall, 1986a; Rambosek and Leach, 1987; Upshall and McKnight, 1987). Recombinant DNA methods as applied to plant pathogens will be discussed in Chapter 9 of this volume.

Exploitation by the pharmaceutical and chemical industries has been considerable. Many species, initially selected because of their ability to produce a commercially important metabolite at significant levels, are involved, and many reviews have been written detailing the reasons for, and the specifics of, this exploitation (e.g., Sermonti, 1969; Ball, 1983; Smith and Berry, 1975, 1976, 1978; Smith et al., 1983). The impact ranges from medicine, with the production of antibiotics, to the food and detergent industries, which use enzymes, such as proteases and amylases, and other metabolites, such as citric acid.

Recently, several articles have been written outlining the rationale for applying molecular methods to filamentous fungal species (Upshall, 1986a; Cullen and Leong, 1986; Rambosek and Leach, 1987; Upshall and McKnight, 1987). Gradually the speculations and predictions that have been made are being realized. For example, the successful cloning of the *cef*EF gene (encoding the ring-expanding enzyme) from *Cephalosporium acremonium* and its reintroduction into a production strain in multiple copies to overcome a rate-limiting step resulted in a transformant that gave a 47% increase in the production of cephalosporin over the parental strain (Cantwell et al., 1988). This strategy overcomes the time-consuming, labor-intensive mutagenesis/screening programs that rely on rare genetic changes. Another example is the expression of a fungal lipase, cloned from a non-commercial species, in a commercially useful *Aspergillus* species (NOVO Industri, Annual Report, 1988). This latter development is especially relevant, since it makes use of strains that have desired properties, in this case secretion and commercial utility, for the production of a previously unavailable enzyme. Exploitation of commercially acceptable genotypes clearly is of immense value to producing companies. Development costs incurred during strain-breeding programs are markedly reduced, and there is a speedy introduction of a new product to market. Speed of introduction is important in the highly competitive industrial enzyme market, where customer acceptance and being first to market are vital criteria for product survival. Examples of this nature will inevitably increase as the technology becomes applied to more diverse species.

The methods used for cloning genes, their manipulation, and the introduction of DNA into organisms are now routine in most molecular biology laboratories. Many reviews covering this literature have been written, and several chapters of this volume will cover aspects of this methodology in detail. This article concentrates on a topic that has received much attention, especially by the smaller biotechnology companies, the development of filamentous fungi as host organisms for the high level expression of commercially valuable mammalian proteins, both therapeutic and industrial. The impetus for these attempts has been primarily to exploit the secretion capability of filamentous fungi with the anticipation that use could be made of some of the higher yielding industrial strains (Upshall, 1986a). The additional reasons are the proven manipulability of the organisms and the perceived need for alternative expression systems to replace or supplement the yeast system, where problems have been encountered in the expression of some proteins (Lemontt et al., 1985; MacKay, 1987), and of mammalian cell systems, which generally suffer from lower yields and expensive media requirements. Three commercial groups embarked on this strategy (in alphabetical order): Allelix of Toronto, Genencor of San Francisco, and ZymoGenetics of Seattle. The publications from each group provide clear insight into their strategies for the molecular manipulation of fungal species and have also generated information on aspects of the biology of these species. Each group initially chose to develop an expression system in *A. nidulans*. This species is not commercial, but it exhibits many features that are present in related industrial species. It secretes proteins, but not at the level of wild-type strains of other species such as *A. niger* (Smith and Berry, 1975, 1976, 1978; Smith et al., 1983); it produces penicillin, but at an extremely low level when compared to *Penicillium chrysogenum* (Ditchburn et al., 1974); and it expresses proteases (Cohen, 1972, 1977). Perhaps the main reason, however, was that at the time when the work was begun, gene-transfer methods had not been developed for industrial species. It was obvious that progress in this area would only be made by exploiting the development of *A. nidulans* technology and by making use of its considerably more advanced genetics and mutant strain pool.

II. HETEROLOGOUS PROTEIN EXPRESSION

There are two basic components to consider in developing a heterologous protein expression/secretion system:

1. transcriptional and translational control of the mammalian DNA, usually a cDNA molecule, and
2. protein processing and modification during its passage through the secretory pathway of the cell.

Additional factors are the resolution of problems inhibiting the achievement of maximum yield, such as the elimination of extracellular proteases. To date, the first two aspects have been approached by molecular means.

A. Transcriptional and Translational Control of Gene Expression

To ensure that the mammalian cDNA molecules would be expressed, each group developed an expression system around a fungal promoter system.

The most straightforward approach was that of Cullen et al. (1987) and Gwynne et al. (1987a), who employed the promoter of the starch-regulated glucoamylase (AMG) gene from *A. niger*. This gene was easy to clone since the complete DNA sequence of both the *A. niger* and *A. awamori* genes were published (Nunberg et al., 1984; Boel et al., 1984).

Working on the hypothesis that fungal promoters would function across species, these workers fused the fungal transcriptional and translational signals to various versions of the prochymosin gene (Cullen et al., 1987) or α interferon (Gwynne et al., 1987a), respectively, and transformed *A. nidulans*. Not only did this promoter function, but both prochymosin and α-interferon production was regulated by starch. Gwynne et al. (1987a) also developed an expression system based on the ethanol-regulated promoter of the *alc*A gene, cloned from *A. nidulans* by Lockington et al. (1986). This promoter is glucose repressed, induced by ethanol or one of several other gratuitous inducers such as threonine, and is also positively regulated by the *alc*R gene, the product of which complexes with the inducer to activate the *alc*A promoter (Pateman et al., 1983). The initial expression plasmid was constructed using convenient restriction enzyme sites and was deleted for 52 nucleotides of the *alc*A untranslated mRNA leader sequence. Although this construction allowed the production of α interferon, restoring the deleted sequence, especially with one that contained the natural eight nucleotides immediately preceding the *alc*A initiating ATG, on average doubled the levels of α interferon produced. However, it was found that, even in transformants carrying this optimized translation initiation region, the α interferon yield was limited by the low-level expression of the single indigenous copy of the positive regulator gene *alc*R. By analogy with a similar effect observed for the regulation of genes controlled by the *amd*R gene (Kelly and Hynes, 1987), it was postulated that the multiple copies of the alcA promoter in high copy-number transformants titrated the positively regulating *alc*R product such that each copy of the promoter was not functioning at maximum efficiency. Introduction of multiple copies of the cloned *alc*R gene into the transformants resulted in significantly increased production of α interferon (Gwynne et al., 1987b).

In our work at ZymoGenetics we have examined the effect of three different promoters on the expression of tissue plasminogen activator (tPA). These promoters were from: (a) an ethanol-regulated but not glucose-repressed alcohol dehydrogenase gene (*adh3*, renamed *alc*C) of *A. nidulans* (McKnight et al., 1985); (b) the triosephosphate isomerase (*tpi*A) gene, a medium-level, constitutively expressed gene from the same species (McKnight et al., 1986); and (c) a high-level, constitutively expressed alcohol dehydrogenase (*adh*A) from *A. niger* (Upshall et al., unpublished). Each of these genes was isolated following complementation of yeast mutants with cDNA libraries made in a yeast expression vector modified from one described in McKnight and McConaughy (1983).

The promoter fragment used in each case was greater than 1 kb in length and carried all necessary transcriptional and translational signals, with the exception of an initiating ATG. The 3' end of each was modified to carry both an *Eco*RI and either a *Bam*HI or *Bgl*II restriction-enzyme site adjacent to and upstream of the natural ATG of each respective gene. In every construction tested, the terminator was from the *tpiA* gene. The promoters were fused to a tPA cDNA modified to carry a *Bam*HI restriction site adjacent to and on the 5' side of the initiating ATG codon, and inserted into pM048, a pBR322-based vector carrying the wild-type allele of the *arg*B gene to allow the selection of transformants (see Figure 1 in Upshall et al., 1987). All constructions were transformed into the same strain and 10 independent randomly selected transformants were grown as 50-ml cultures in 250-ml shake flasks for 2 days at 37 °C. Sampled medium was filtered through 0.2-micron membrane filters and directly tested for secreted levels of tPA in fibrin lysis plates, as described in Upshall et al. (1987). The mean and range of titers from each set of strains is shown Table 1. The highest titer was obtained from the transformants carrying the *adh*A promoter/tPA cDNA fusions. Transformants carrying the *alc*C promoter/tPA fusion gave the lowest yield, even when grown in the presence of ethanol.

Table 1 tPA Titers Using Different Promoters

Construction	Mean titer μg/l	Titer range μg/l
*tpi*A/tPA	12.5	3.2-47
*adh*A/tPA	99.3	13-300
*alc*C/tPA (+ eth)	3.5	0-7.2
*adh*A/amg/tPA	46.2	8-125
*alc*C/amg/tPA (+ eth)	1.3	0-3.4

Titers obtained with either the natural tPA secretory signals or with constructions in which these signals were replaced by those of the *A. niger* glucoamylase gene. The transformants carrying the *alc*C promoter constructions were grown in the presence of 1% ethanol. All cultures were grown for 28 hr at 37 °C under conditions described in Ref. 29. The range of titers is that observed for 10 independent transformants.

In all of these studies, a recurring feature is the variability of expression levels observed with independent transformants. In filamentous fungi, donor DNA becomes integrated into the chromosome at either a homologous site, as it does in about 80% of the cases in *A. nidulans* with *arg*B as the selective marker (Upshall, 1986b), or in nonhomologous locations, as in the majority of cases with *pyr-4*-selected transformants (Ballance et al., 1983). Part of the variation in titer could be the result of differences in expression because of a "chromosome context" effect at the site of integration. Some chromosomal regions may allow better expression than others. An alternative explanation is that expression is correlated with the number of copies of integrated plasmid. This appears to be the case for the ethanol-regulated *alc*A gene (Gwynne, 1987a), but in our work we could find no significant relationship between copy number and titer. A similar picture emerged in the analysis of Cullen et al. (1987), where a transformant with fewer integrated copies secreted the most prochymosin. We have not comprehensively investigated mRNA levels in these transformants.

B. Protein Processing and Modification

One of the major reasons for developing filamentous fungi as expression hosts is to take advantage of their prodigious secretion capability. The primary translation product of secreted proteins generally has a hydrophobic signal sequence and sometimes a short leader peptide at the N terminus, neither of which appear in the mature protein. These signal sequences are cleaved from the translation product by enzymes that recognize the specific amino acid sequences that identify cleavage sites. These processing enzymes have not yet been isolated and characterized from filamentous species, but a good example is the enzyme encoded by the *KEX2* gene of *S. cerevisiae*, which is required for processing of the mating pheromone α factor (Fuller et al., 1985) and killer toxin (Thomas et al., 1987). That such enzymes exist in the filaments is shown by the molecular analysis of the glucoamylase gene (Nunberg et al., 1984; Boel et al., 1984). By comparing the N-terminal amino acid sequence of purified secreted protein with the translated cDNA sequence, the processing sites within the amino terminal peptide could be determined. Two sites have been resolved, cleavage at the first removes the hydrophobic signal (prepeptide) and at the second, a hexapeptide leader (prosequence) in which the two carboxy-terminal amino acids are Lys-Arg. The same pair of basic amino acids are also at a predicted cleavage site for the processing of one of the ligninase enzymes encoded by a gene isolated from *Phanerochaete chrysosporium*

(Tien and Tu, 1987). Secreted proteins from other fungal species have been shown to have different processing signals (e.g., Penttila et al., 1986; Teeri et al., 1987). The efficiency of the natural versus a fungal secretory signal in directing secretion and processing have been compared for pro-chymosin and tPA. Cullen et al. (1987) employed three constructions: one in which the AMG (glucoamylase) signal peptide was fused to pro-chymosin; a second in which the full AMG prepro sequence (signal and hexapeptide leader) was fused to prochymosin; and a third that carried the natural, full-length preprochymosin. It was found that, on average, transformants carrying the AMG signal-prochymosin fusion gave higher yields than transformants carrying the natural, full-length preprochymo-sin. Transformants carrying the AMG prepro-prochymosin fusion gave the lowest yield. These results suggested that, although *A. nidulans* pro-cessing enzymes were able to recognize the natural mammalian signal, their cleavage may not be as efficient as in the molecule carrying the AMG signals. An alternative explanation is that the mRNA encoding prepro-chymosin may not be translated with the same efficiency as that from the AMG-prochymosin fusion.

We have performed a similar study for tPA expression. The primary transcript of tPA has both a signal and a leader peptide (prepro form), and amino-terminal amino acid analysis of purified mammalian protein has identified two forms. The first results from cleavage after an Arg-Arg pair of amino acids to create the glycine-form, which, following amino-peptidase removal of four amino acids, yields the serine form (Pennica et al., 1983). We have obtained transformants carrying either a fusion of DNA coding for the AMG prepro peptide directly to the DNA encoding the glycine form of tPA, or the natural, full-length tPA cDNA. The mean level of tPA obtained with the fusion protein was significantly less than the mean level obtained from transformants carrying the natural tPA cDNA molecule (Table 1). This result is opposite to the one obtained for prochymosin, but we have no understanding of the underlying reasons. One possible explanation is that processing is related to the conforma-tion of the molecule. In the case of the AMG/tPA fusion, the primary translation product may not have assumed the correct conformation to present the processing site for proteolytic cleavage, but as with prochy-mosin, efficiency of mRNA translation is an alternative explanation.

The accuracy of the proteolytic processing, an important component of heterologous protein expression studies, can be determined by amino-terminal amino acid sequence analysis of the secreted protein.

We have performed this analysis on tPA obtained from cultures grown under two different conditions. In the first, performed on protein secreted from an *A. nidulans* transformant carrying full-length tPA cDNA under the control of the *tpi*A promoter, it was found that both the naturally occurring serine and glycine forms were produced. In this experiment the yield of tPA was approximately 100 μg/l from 28-hr-old cultures (Upshall et al., 1987). In a subsequent experiment, tPA was purified from the medium of a 4-day-old culture of a transformant carrying the natural tPA cDNA transcriptionally controlled by the highly expressed *adh*A promoter sequence. The yield of tPA was 2.5 mg/l and amino-terminal amino acid analysis identified only the serine form of the protein. The results of these two experiments clearly show that *A. nidulans* has proteolytic processing enzymes that recognize some mammalian signals and cleave efficiently with the desired specificity. This is of considerable importance in determining the choice of an expression host and gives filamentous fungi a significant advantage over other microbial systems for some proteins.

However, it became of interest to determine the versatility of *A. nidulans* in terms of its abilities to both express and correctly process mammalian proteins, the processing signals of which are different from tPA. We used human granulocyte-macrophage colony-stimulating factor (h-GMCSF), which has a nominal molecular weight of 16 kD and is modified with both O- and N-linked carbohydrate (Kaushansky et al., 1986). The amino acid sequence around the signal-peptide processing site is:

. . . . . . gly-thr-val-ala-cys-ser-ile-ser-ala-pro-ala . . . .

↑

site of normal cleavage

Full-length cDNAs encoding three forms, the natural, one lacking N-linked carbohydrates, and one totally deficient in potential carbohydrate-binding sites, were fused to the *adh*A promoter sequence and plasmid constructions transformed into *A. nidulans*. Between 500 μg and 1 mg per liter of active protein were obtained from the highest yielding transformants in each case. Amino acid analysis of the mutant, carbohydrate-deficient form of the protein revealed two amino-terminal sequences, neither of which corresponded to the expected amino terminus (Table 2). No forms with amino termini that would be indicative of processing upstream of the cleavage sites were found. Clearly, all of the amino-terminal signal sequences had been removed from the secreted material, but it is not known whether the observed amino-terminal heterogeneity is a consequence of aminopeptidase activity following correct proteolytic cleavage or whether the fungus harbors a protease that preferentially cleaves within Ala-Pro-rich regions. We have not yet determined the amino-terminal sequence of the secreted natural form of the molecule.

Table 2 Amino-Terminal Amino Acid Analysis of the Carbohydrate Free Form of GMCSF Expressed from *A. nidulans*

Cycle	Observed 16%	Observed 84%	Expected
1	Ala-3	Pro-6	Ala-1
2	Arg	Ala	Pro
3	Ala	Pro	Ala
4	Pro	Ala	Arg
5	Ala	Thr	Ala
6	Pro	Glu	Pro
7	Ala	Pro	Ala
8	Thr	Try	Pro
9	Glu	Gln	Ala
10	Pro	His	Thr
11	Try	Val	Glu
12	Gln	Asn	Pro
13	His	Ala	Try
14	Val	Ile	Gln
15	Asn	Glu	His

In addition to proteolytic processing, most secreted proteins are glycosylated by the addition of carbohydrate. There are two types of glycosylation: N linked, in which the sugars are attached to an asparagine-linked N-acetyl-d-glucosamine backbone, and O linked, in which the carbohydrate is added to serine or threonine in regions of the protein rich in these amino acids. The precise function of the added carbohydrate is unknown, but its removal can markedly alter the properties of the protein (Kaushansky et al., 1987). Fungal proteins can also be glycosylated with both N- and O-linked additions (Nunberg et al., 1984; Boel et al., 1984; Salovuori et al., 1987). Although it has been shown that both the pattern of addition and the structure of the added carbohydrate is similar to that in higher eukaryotic cells (Wasserman and Hultin, 1981; Salovouri et al., 1987), the nature and quantity of the added carbohydrate is important, especially with therapeutic proteins, because of its antigenic properties. Hyperglycosylation, the addition of large numbers of mannose sugars to the N-acetyl glucosamine/mannose core backbone, has been a major problem with the expression of some proteins from yeast (Lemontt et al., 1985; MacKay, 1987). In our analysis of the tPA produced from *A. nidulans* transformants carrying the *tpi*A promoter/tPA cDNA expression unit (levels of 100 μg/l), the secreted protein comigrated with authentic mammalian tPA in SDS-PAGE gels (Upshall et al., 1987), indicating that there was no excess glycosylation. However, gel analysis of purified protein obtained from transformants carrying the *adh*A promoter/tPA cDNA expression unit (levels of 2.5 mg/l) showed this protein to carry excess carbohydrate. This was the same protein sample that was correctly processed to generate only the serine form, as detailed above. We have detected similar hyperglycosylation of mammalian alpha-1-antitrypsin when expressed at 1.5 mg/l in 3-day-old cultures of transformed strains. In this case also, the secretory signal was removed. An interpretation of these data is that in the high-yielding transformants the translation of mRNA into protein proceeds faster than the ability of the cells to process and passage the molecule through the secretory apparatus. The intracellular accumulation of protein, coupled with its slower transport, presumably partly in the golgi apparatus, will cause the continued addition of sugar side chains. Clearly, in *A. nidulans* the addition of this extra carbohydrate does not interfere with the proteolytic processing of the secretory signals. We have not analyzed the nature or structure of the carbohydrate added to the proteins that we have produced. Overglycosylation of an aspartyl protease derived from *Mucor mehei* expressed in *A. nidulans* (Gray et al., 1986) and *Aspergillus oryzae* (Christensen et al., 1988) has also been observed.

III. CONCLUSIONS AND PERSPECTIVE

Overall these results indicate that *A. nidulans*, and presumably related fungi that secrete at higher levels, have the cellular capacity to accurately process some mammalian proteins. The problem of hyperglycosylation is one that can be resolved by genetic and molecular manipulation, especially by cloning genes involved in the secretion and glycosylation processes, as has been done for *S. cerevisiae* (Smith et al., 1985; Yip et al., 1988). The problem of proteases can be similarly resolved. In our work on expression of tPA and GMCSF we have minimized protease activity by maintaining the pH of the culture medium between 6.5 and 7.0; by using a strain mutant in the *are*A gene, the regulatory gene controlling the expression of genes involved in nitrogen metabolism, including some proteases (Arst and Cove, 1973); and by incorporating a high concentration of ammonium ions into the culture medium. However, we have observed proteolytic degradation if the pH falls below 6.0, indicating the presence of either acid-activated or acid-induced proteases that are not subject to the nitrogen regulatory pathway. Many proteases have been identified by several groups (Cohen, 1972, 1977). Purification and amino acid sequence analysis of these should present no problem with current methods. From derived amino acid sequences, oligonucleotides can be made and used as hybridization probes to isolate the genes. Although gene-disruption and gene-replacement methods are not as efficient as in *S. cerevisiae*, such genetic manipulation has been reported for several species (Rambosek and Leach, 1987; Miller et al., 1985; Paietta and Marzluf, 1985). This means of eliminating proteases will become routine and an integral part of strain development programs.

Filamentous fungi clearly have the essential components necessary for their use as a system for the expression of some heterologous proteins. However, their acceptance for this purpose will depend on the enthusiasm of their protagonists and the continued financial support for a molecular involvement in strain development.

ACKNOWLEDGMENTS

For excellent technical assistance we wish to thank S. Osborne, L. deJong, and K.P.L. Lewison.

REFERENCES

Arst, H.N. and Cove, D.J. (1973). Nitrogen metabolite repression in *Aspergillus nidulans*. *Mol. Gen. Genet. 126*: 111-142.

Ball, C. (1983). Filamentous fungi. In *Genetics and Breeding of Industrial Microorganisms*, Ball, C., ed., CRC Press, Boca Raton, FL, p. 159-188.

Ballance, D.J., Buxton, F.P., and Turner, G. (1983). Transformation of *Aspergillus nidulans* by the orotidine-5'-phosphate decarboxylase gene of *Neurospora crassa*. *Biochem. Biophys. Res. Commun. 112*: 284-289.

Boel, E., Hjort, I., Svennson, B., Norris, F., Norris, K.E., and Fiil, N.P. (1984). Glucoamylases G1 and G2 from *Aspergillus niger* are synthesized from two different but closely related mRNAs. *EMBO J. 3*: 1097-1102.

Cantwell, C., Skatrud, P., Chapman, J., Dotzlaf, J., Yeh, W.K., Fisher, T., Samson, S., Ingolia, T.D., and Queener, S.W. (1988). Biosynthetic studies with a recombinant strain of *Cephalosporium acremonium*. *S.I.M. News 38* (Suppl): 22.

Case, M.E., Schweizer, M., Kushner, S.R., and Giles, N.H. (1979). Efficient transformation of *Neurospora crassa* by utilizing hybrid plasmid DNA. *Proc. Natl. Acad. Sci. USA 76*: 5259-5363.

Christensen, T., Woeldike, H., Boel, E., Mortensen, S.B., Hjortshoej, K., Thim, L., and Hansen, M.T. (1988). High level expression of recombinant genes in *Aspergillus oryzae*. *Bio/Technology 6*: 1419-1422.

Cohen, B.L. (1972). Ammonium repression of extracellular protease. *J. Gen. Microbiol. 71*: 293-299.

Cohen, B.L. (1977). The proteases of *Aspergilli*. In *The Genetics and Physiology of Aspergillus*, Smith, J.E. and Pateman, J.A., eds., Academic Press, London, p. 281-292.

Cullen, D., Gray, G.L., Wilson, L.J., Hayenga, K.J., Lamsa, M.H., Rey, M.W., Norton, S., and Berka, R.M. (1987). Controlled expression and secretion of bovine chymosin in *Aspergillus nidulans*. *Bio/Technology 5*: 369-376.

Cullen, D. and Leong, S. (1986). Recent advances in the molecular genetics of industrial filamentous fungi. *Trends Biotechnol. 4*: 285-288.

Ditchburn, P., Giddings, B., and MacDonald, K.D. (1974). Rapid screening for the isolation of mutants of *Aspergillus nidulans* with increased penicillin yields. *J. Appl. Bacteriol. 37*: 515-523.

Fuller, R.S., Brake, A.J., Julius, D.J., and Thorner, J. (1985). The KEX2 gene product required for processing of yeast prepro-alpha-factor is a calpain like endopeptidase specific for cleaving at pairs of dibasic residues. In *Protein Transport and Secretion*, Gething, M.J., ed., Cold Spring Harbor Laboratory, Cold Spring Harbor, NY, p. 95.

Gray, G.L., Hayenga, K., Cullen, D., Wilson, L.J., and Norton, S. (19867). Primary structure of *Mucor mehei* aspartyl protease: evidence for a zymogen intermediate. *Gene 48*: 41-53.

Gwynne, D.I., Buxton, F.P., Williams, S.A., Garven, S., and Davis, R.W. (1987a). Genetically engineered secretion of human interferon and a bacterial endoglucanase from *Aspergillus nidulans*. *Bio/Technology 5*: 713-719.

Gwynne, D.I., Buxton, F.P., Gleeson, M.A., and Davis, R.W. (1987b). Genetically engineered secretion of foreign proteins from *Aspergillus nidulans*. In *Protein Purification: Micro to Macro*, Burgess, R., ed., Alan R. Liss, New York, p. 355-365.

Kaushansky, K., O'Hara, P.J., Berkner, K., Segal, G.M., Hagen, F.S., and Adamson, J.W. (1986). Genomic cloning, characterization, and multilineage growth promoting activity of human granulocyte macrophage colony-stimulating factor. *Proc. Natl. Acad. Sci. USA 83*: 3101-3105.

Kaushansky, K., O'Hara, P.J., Hart, C.E., Forstrom, J.W., and Hagen, F.S. (1987). Role of carbohydrate in the function of human granulocyte-macrophage colony-stimulating factor. *Biochemistry 26*: 4861-4867.

Kelly, J.M. and Hynes, M.J. (1987). Multiple copies of the *amd*S gene of *Aspergillus nidulans* cause titration of trans-acting regulatory proteins. *Curr. Genet. 12*: 21-31.

Lemontt, J.F., Wei, C.M., and Dackowski, W.R. (1985). Expression of human type tissue plasminogen activator in yeast. *DNA 4*: 419-428.

Lockington, R.A., Sealy-Lewis, H.M., Scazzochio, C., and Davies, R.W. (1986). Cloning and characterization of the ethanol utilization regulon in *Aspergillus nidulans*. *Gene 33*: 137-149.

MacKay, V.L. (1987). Secretion of heterologous proteins in yeast. In *Biochemistry and Molecular Biology of Industrial Yeasts*, Stewart, G.G., Russell, I., Klein, R.D., and Hiebsch, R.R., eds., CRC Press, Boca Raton, FL, p. 27-36.

McKnight, G.L. and McConaughy, B.L. (1983). Selection of functional cDNAs by complementation in yeast. *Proc. Natl. Acad. Sci. USA 80*: 4412-4416.

McKnight, G.L., Kato, H., Upshall, A., Parker, M.D., Saari, G., and O'Hara, P.J. (1985). Identification and molecular analysis of a third *Aspergillus nidulans* alcohol dehydrogenase gene. *EMBO J. 4*: 2093-2099.

McKnight, G.L., O'Hara, P.J., and Parker, M.L. (1986). Nucleotide sequence of the triosephosphate isomerase gene from *Aspergillus nidulans*: implications for a differential loss of introns. *Cell 46*: 143-147.

Miller, B.L., Miller, K.Y., and Timberlake, W.E. (1985). Direct and indirect gene replacements in *Aspergillus nidulans*. *Mol. Cell. Biol. 5*: 1714-1721.

Novo Industri, *Annual Report 1988*, Novo-Nordisk A/S, Novo Alle 2800, Bagsvaerd Denmark.

Nunberg, J.H., Meade, J.H., Cole, G., Lawyer, F.C., McCabe, P., Schweickart, V., Tal, R., Wittman, V.P., Flatgaard, J.E., and Innes, M.A. (1984). Molecular cloning and characterization of the glucoamylase gene of *Aspergillus awamori*. *Mol. Cell. Biol. 4*: 2306-2315.

Paietta, J.V. and Marzluf, G.A. (1985). Gene disruption by transformation in *Neurospora crassa*. *Mol. Cell. Biol. 5*: 1554-1559.

Pateman, J.A., Doy, C.H., Olsen, J.E., Norris, U., Creaser, E.H., and Hynes, M.J. (1983). Regulation of alcohol dehydrogenase (ADH) and aldehyde dehydrogenase (AldDH) in *Aspergillus nidulans*. *Proc. R. Soc. Lond. B217*: 243-264.

Pennica, D., Holmes, W.E., Kohr, W.E., Hawkins, R.N., Vehar, G.A., Ward, C.A., Bennett, W.F., Yelverton, E., Seeburg, P.H., Heynecker, H.L., and Goeddel, D.V. (1983). Cloning and expression of human tissue type plasminogen activator cDNA in *E. coli*. *Nature 301*: 214-221.

Penttila, M., Lehtovaara, P., Nevalainen, H., Bhikhabhia, R., and Knowles, J. (1986). Homology between cellulose genes of *Trichoderma reesei*: complete nucleotide sequence of the endoglucanase gene. *Gene 45*: 253-263.

Rambosek, J. and Leach, J. (1987). Recombinant DNA in filamentous fungi: progress and prospects. In *Critical Reviews in Biotechnology*, Vol. 6, Stewart, G.G. and Russell, I., eds., CRC Press, Boca Raton, FL, p. 357-393.

Salovuori, I., Makarow, M., Rauvala, H., Knowles, J., and Kaarainen, L. (1987). Low molecular weight high-mannose type glycans in a secreted protein of the filamentous fungus *Trichoderma reesei*. *Bio/Technology 5*: 152-156.

Sermonti, G. (1969). *Genetics of Antibiotic Producing Micro-Organisms*. Wiley, London.

Smith, J.E. and Berry, D.R. (1975). *The Filamentous Fungi*, Vol. 1. Edward Arnold, London.

Smith, J.E. and Berry, D.R. (1976). *The Filamentous Fungi*, Vol. 2. Edward Arnold, London.

Smith, J.E. and Berry, D.R. (1978). *The Filamentous Fungi*, Vol. 3. Edward Arnold, London.

Smith, J.E., Berry, D.R., and Kristianson, B. (1983). *The Filamentous Fungi*, Vol. 4. Edward Arnold, London.

Smith, R.A., Duncan, M.J., and Moir, D.T. (1985). Heterologous protein secretion from yeast. *Science 229*: 1219-1224.

Teeri, T.T., Lehtovaara, P., Kauppinen, S., Salovuori, I., and Knowles, J. (1987). Homologous domains in *Trichoderma reesei* cellulolytic enzymes: gene sequence and expression of cellobiohydrolase II. *Gene 51*: 43-52.

Thomas, D.Y., Vernet, T., Boone, C., Greene, D., Lolle, S., and Bussey, H. (1987). The yeast killer toxin expression system: functional analysis and biotechnological applications. In *Biological Research on Industrial Yeasts*, Vol. I, Stewart, G.G., Russell, I., Klein, R.D., and Heibsch, R.R., eds., CRC Press, Boca Raton, FL, p. 155-163.

Tien, M. and Tu, C.-P.D. (1987). Cloning and sequencing of a cDNA for a ligninase from *Phanerochaete chrysosporum*. *Nature 326*: 520-523.

Upshall, A. (1986a). Filamentous fungi in biotechnology. *Bio/Techniques 4*: 158-166.

Upshall, A. (1986b). Genetic and molecular characterization of *arg*B transformants of *Aspergillus nidulans*. *Curr. Genet. 10*: 593-599.

Upshall, A., Kumar, A.A., Bailey, M.C., Parker, M.D., Favreau, M.A., Lewison, K.P.L., Joseph, M.L., Maraganore, J.M., and McKnight, G.L. (1987). Secretion of active human tissue plasminogen activator from the filamentous fungus *Aspergillus nidulans*. *Bio/Technology 5*: 1301-1304.

Upshall, A. and McKnight, G.L. (1987). Filamentous fungi: hosts for the expression of foreign proteins. In *Biological Research on Industrial Yeasts*, Vol. I. Stewart, G.G., Russell, I., Klein, R.D., and Heibsch, R.R., eds., CRC Press, Boca Raton, FL, p. 129-136.

Wasserman, B.P. and Hultin, H.O. (1981). Effect of deglycosylation on the stability of *Aspergillus niger* catalase. *Arch. Biochem. Biophys. 212*: 385-392.

Yip, C.L., Welch, S.K., Gilbert, T., and MacKay, V.L. (1988). Cloning and sequencing of the *S. cerevisiae* MNN9 gene required for hyperglycosylation of secreted proteins. *Yeast 4* (Special Issue): S457.

3
Molecular Biology of a High-Level Recombinant Protein Production System in *Aspergillus*

R. Wayne Davies
University of Glasgow, Glasgow, Scotland

I. INTRODUCTION

In this chapter the development, status, and industrial use of *Aspergillus nidulans* and *A. niger* expression-secretion systems for recombinant proteins will be described, with particular emphasis on a system based on the ethanol utilization regulon of *A. nidulans*. In the work described, a laboratory strain that was not preselected for any parameter of industrial performance has been converted by the application of relatively simple molecular biology into a competitive industrial production system.

The primary goal of the program described here was to achieve a commercially viable position in the production of protein products from recombinant DNA sources by relying on the natural propensity of filamentous fungi to secrete large amounts of enzymes into the extracellular medium. Previous approaches to utilizing this natural resource were limited to the generation of specialized enzyme-producing strains by screening wild isolates and by mutagenesis and screening. Each of the resulting strains is very specialized and has a genetic history that cannot be determined or recapitulated. In contrast, we set out to develop a set of strains with a precisely

45

defined genetic constitution and history (one particular strain underlies all genetic work in *A. nidulans*) and with the flexibility to produce any protein. It was clear in 1982 that even a modest application of modern molecular technology to filamentous fungal species would, with a high probability, allow the creation of a competitive protein production system—how competitive could only be determined in practice.

Among the species considered as hosts for the development of this system, *A. niger* stood out as being in particularly widespread use in the enzyme industry (Barbesgaard, 1977) and had GRAS (generally regarded as safe) status in the United States for food industry products; however, no transformation system existed, and sexual genetics is impossible. A related species, *A. nidulans*, had been the subject of investigation by classical and molecular geneticists for many years, and presented a well-developed and convenient genetic system. In *A. nidulans* a number of regulatory systems involving primarily catabolic pathways of nitrogen (and to a lesser extent carbon) metabolism, e.g., purine degradation, nitrite and nitrate utilization, acetamide utilization, and ethanol utilization, had been excellently characterized by classical genetics. The number and identity of structural genes of these and other systems had been determined, and the basal and induced levels of enzyme activity, and in many cases the actual level of enzyme protein production, were known. *Trans*-acting controlling factors had been identified genetically and had been formally shown to be activator (or more rarely repressor) elements. Transformation systems had been developed (Tilburn et al., 1983; Ballance et al., 1983) and advanced techniques such as specific gene replacement were under development. Thus the analysis of genetically engineered strains was clearly facilitated by using *A. nidulans*, and all the tools were available for research aimed at increasingly advanced tuning of an expression system. For these reasons and despite the unproven capacity of this species to produce high levels of proteins in fermentors, the primary expression-secretion system development was carried out in *A. nidulans*. In parallel, adequate technology was developed to allow the exploitation of *A. niger* as a host, should that prove desirable. The continuing impact on the development of expression systems in *E. coli* of ever-increasing genetic and biochemical knowledge of plasmid replication and segregation, transcription, and translation mechanisms, the secretion pathway, etc. is sufficient demonstration of the importance of working with a system where goal-oriented basic research can be carried out efficiently. In fact, *A. nidulans* and *A. niger* do seem to be sufficiently closely related for technology to be transferred from one to the other with minor modifications. Table 1 summarizes the reasons for the choice of this system.

Table 1 Practical Advantages of the *Aspergillus alcA* System

General	*alcA*	Secretion
Aspergillus		
High natural level of protein secretion	Strong inducible promoter	Simple purification
Eukaryotic	Very low basal level	High initial purity
S-S bonds formed well	Cheap gratuitious inducers available	Continuous fermentation possible
Glycosylation and other posttranslational modifications	1 hr to maximum protein synthesis after induction	Avoids problems of insolubility and cytoplasmic instability of protein products
Rapid growth on inexpensive media	mRNA not unstable	Avoids intracellular accumulation of toxic products
Mitotically stable multicopy tansformants	Natural genes dispensable	Sequence-specific endoproteases in pathway
A. nidulans		
Excellent genetics	Simple, defined control system	Takes advantage of posttranslational modifications
Molecular biology similar to yeast	Control gene cloned and sequenced	Allows production of domains of proteins
Transformation, gene replacement established	Structural genes cloned and sequenced	
A. niger		
F.D.A. approved	Fermentation procedures established	
Established commercial enzyme-producing strains		
Fermentation practice established		

The stages in the development of the *Aspergillus* expression-secretion system were:

1. characterization and testing of the necessary molecular components for an *A. nidulans* expression-secretion system and their assembly with recombinant genes coding for foreign proteins;
2. creation of a second-stage prototype production-level system via improvements requiring no preliminary basic research;
3. establishment of the components of a system allowing access to *A. niger* industrial strains and the transfer of *A. nidulans* technology to *A. niger*;
4. ongoing basic research support for the continued improvement of the systems in use, with emphasis on transcription, translation, and secretion signals and mechanisms.

II. THE ETHANOL REGULON OF *A. NIDULANS*

The primary expression-secretion system that was developed is based on the properties of the genetic elements controlling the expression of the ethanol regulon of *A. nidulans* and the involvement of this regulon in the general physiology of the organism. The ethanol regulon of *A. nidulans* (Figure 1) is very simple, comprising two structural genes and one regulatory

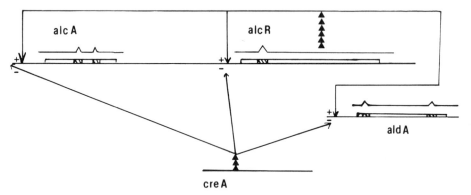

Figure 1 The ethanol regulon of *Aspergillus nidulans*. The most probable situation is given based on evidence available in August 1989. The *alcA, aldA,* and *alcR* genes and transcripts are to scale; *creA* has not yet been characterized. − and + refer to negative and positive control. Introns are indented in mRNAs and cross-hatched in genes.

gene (Pateman et al., 1983; Sealy-Lewis and Lockington, 1984). The structural genes are *alcA*, which codes for alcohol dehydrogenase I (ADH), and *aldA*, encoding aldehyde dehydrogenase (AldDH). The product of the regulatory gene *alcR* acts as a positive activator of both the *alcA* and *aldA* genes and is closely linked to the *alcA* gene on chromosome VII (Figure 1). *aldA* is on chromosome VIII. The physiological inducer of this regulon has not been formally identified, but is probably acetaldehyde. A variety of cheap gratuitous inducers are used in fermentation practice, such as 50 mM ethylmethylketone (Creaser et al., 1984), while in the laboratory it is convenient to use 100 mM threonine (Lockington et al., 1985). All three genes in this regulon are directly and/or indirectly subject to carbon catabolite repression control (Arst, 1981) mediated by the negative control gene *creA* (Bailey and Arst, 1975) (Figure 2).

Under certain induction conditions, both *alcA* and *aldA* mRNAs become major mRNA species, reaching levels that have been reported to be as high as 5% of total mRNA, but are actually around 1%. ADH and AldDH polypeptides are the major in-vitro translation products of RNA extracted from threonine-induced mycelium (Lockington et al., 1985). When uninduced in the presence of acetate or glucose, however, no detectable ADH or AldDH in-vitro translatable mRNA is found (Pateman et al., 1983; Sealy-Lewis and Lockington, 1984; Lockington et al., 1985), and mRNA from these genes is not detectable by radioactive probing of

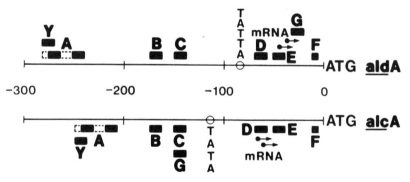

Figure 2   Conserved sequences in the promoter regions of the *alcA* and *aldA* genes. The DNA sequences themselves are given in Gwynne et al. (1987a), where the same alphabetical labels are used, except for Y, which is a sequence with homology to the binding site of ADRI in the yeast ADH2 promoter. Sequences upstream of C can be deleted without affecting *alcR* inducibility; the presence of the right or left half of C is sufficient for *alcR* to act (M. Devchand, personal communication).

Northern transfers (Lockington et al., 1985). Recent experiments in which β-tubulin genes of *A. nidulans* were expressed under the control of the *alcA* promoter (Waring et al., 1989) did allow the detection of a measurable uninduced non-glucose-repressed basal level of *alcA* expression, and even of the glucose-repressed basal level, which is obviously very low, by the occurrence of sensitivity to benomyl in otherwise benomyl-resistant host strains. Nevertheless, it is clear that the *alcA* and *aldA* genes show very high induction ratios and reasonably high maximal levels of mRNA synthesis and mRNA steady-state levels. The stability and mechanism of degradation of these mRNAs has not been investigated; all that can be said is that under normal laboratory conditions they do not appear to be particularly unstable. One feature of this regulon that should be remembered is that the product of the ADH-catalyzed reaction, acetaldehyde, is lethal to the cell if allowed to accumulate. Thus, although the *alcA* and *aldA* genes are coordinately regulated, the system must be preset so that AldDH activity levels are always higher than ADHI levels. The *aldA* promoter does seem on preliminary investigation to be somewhat stronger than the *alcA* promoter, which was actually used in this work because of the convenient occurrence of restriction sites; possible differences in the efficiency of translation and the rate of degradation of mRNA or protein have not been investigated.

The genes of the ethanol regulon were among the few genes in control systems of *A. nidulans* to be cloned before a transformation system was developed. *AlcA* and *aldA* were cloned (Lockington et al., 1985) by using RNA from threonine-induced mycelium in the presence of a 100-fold excess of unlabeled equivalent RNA from an *alcA-alcR* deletion strain as a probe on cDNA libraries made from size-fractionated mRNA. *AlcR* was obtained together with *alcA* in a 13-kb *Eco*RI fragment, the two genes being separated by 2.2 kb. Subsequently the complete nucleotide sequences of all three genes were determined (Gwynne et al., 1987a; Pickett et al., 1987; Felenbok et al., 1988). Comparison of the 5' flanking sequences of the *alcA* and *aldA* genes revealed a series of weakly related sequences that occur in the same order (Figure 2). The same pattern was unfortunately not apparent when the *alcR* promoter (which is also regulated by the *alcR* protein, see below) was sequenced (Felenbok et al., 1988). All three promoters have only one continuous sequence in common, although a number of similarities turn up on pairwise comparison. Dissection of the roles of the various sequence elements in the *alcA* and *aldA* promoters by deletion analysis showed that sequence C in Figure 2 is the *alcR* binding site (M. Devchand, personal communication), with the complication that both 5' and 3' halves of this region can independently direct wild-type *alcR*-dependent expression, although there is no obviously related sequence in the two halves.

The *alcR* gene is known to code for a positive activator, since loss of function mutations in *alcR* result in noninducibility of *alcA* and *aldA* (Pateman et al., 1983; Sealy-Lewis and Lockington, 1984). Nonleaky *alcR* mutations were shown to result in a lack of translatable mRNA (Pateman et al., 1983) and indeed in a lack of mRNA transcription from *alcA* and *aldA* (Lockington et al., 1985), as measured by hybridization. The *alcR* gene itself is autoregulated by the *alcR* protein (Lockington et al., 1987); the level of *alcR* mRNA is inducible some 5- to 10-fold in the presence of ethylmethylketone (Figure 3), and a nonleaky mutation in *alcR* eliminated this inducibility, though not the basal level of *alcR* mRNA synthesis. As with *alcR* and *aldA*, the basal level of the *alcR* message is strongly repressed in the presence of glucose, and induction is prevented. This glucose effect is markedly reduced if the same experiments are carried out in a strain carrying the $creA^d1$ mutation, which is partially defective in carbon catabolite repression (Bailey and arst, 1975) (Figure 3). Thus the negative control by *creA* appears to be epistatic to *alcR* activation, and the basal level of these promoters is set by general cellular transcription factors and *creA* if it affects them.

Until recently it was not clear whether the *creA* protein interacts with all the genes of the ethanol regulon or only with *alcR*. The evidence mentioned above shows that *creA* affects *alcR* directly, and in principle controlling *alcR* is sufficient to control the whole regulon. The *alcA* promoter fragments used in all the commercial work to date have at least 320 nucleotides upstream of the translation initiation codon. The 320-nucleotide region upstream of the ATG has been described as carrying all the *cis*-acting elements necessary for gene expression (Pickett et al., 1987). In fact, as long as the wild-type *alcR* gene was the source of *alcR*, we could not tell whether the *alcA* promoter was subject to direct as well as indirect glucose repression. Preliminary experiments in which *alcR* was driven by the constitutive G6PD promoter (F. P. Buxton, D. I. Gwynne, and R. W. Davies, unpublished) indicated that *alcA* was inducible in the presence of glucose in such strains; in addition, there is evidence (M. Devchand, personal communication) for a site of *creA*-mediated carbon catabolite repression of *alcA* located upstream of the fragment containing the sequences necessary for *alcR*-activated transcription in the *alcA* promoter. It seems likely that *alcA* is subject to both direct and indirect glucose repression. This makes physiological sense as a fail-safe mechanism, since

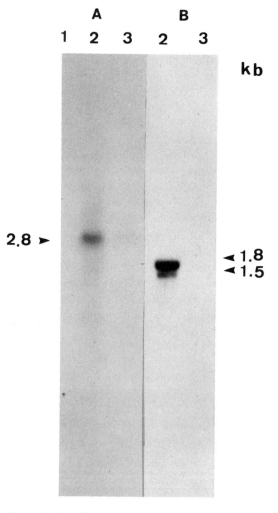

Figure 3 Regulation of the transcripts of the genes of the ethanol regulon. (a) Regulation on a wild-type background. Poly A⁺ RNA was extracted from wild-type mycelium (*biA1*) grown under the following conditions: 1 = carbon catabolite repressed conditions: 0.1% glucose and 50 mM ethylmethylketone; 2 = induced conditions: 0.1% fructose and 50 mM ethylmethylketone; 3 = noninduced conditions: 0.1% fructose. 10 g of poly A⁺ was electrophoretically separated on a 0.75% agarose formaldehyde gel, blotted onto nitrocellulose, and hybridized with nick-translated probes. Panel A: DNA of plasmid pAN105 (Lockington et al., 1985) carrying the *alcR* sequence. Panel B: DNA of plasmid pAN204 carrying the *alcA* sequence and of plasmid pAN210 (Pickett et al., 1987) carrying the *aldA* sequence.

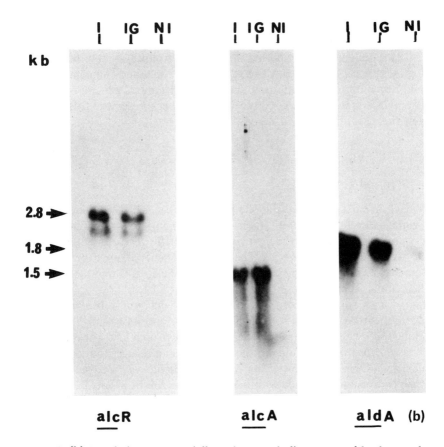

Figure 3 **(b)** Regulation on a partially carbon-catabolite-repressed background. Poly A$^+$ RNA was extracted from a *creA*^{d}1 mutant strain (reduced function of creA; Bailey and Arst, 1975) in induced conditions (I), non-induced conditions (NI), and glucose-repressed conditions (IG), 10 μg fractionated on an 0.75% agarose formaldehyde gel and hybridized. The nick-translated probes used were the same as in Figure 3a.

the product of basal level activity of ADHI is toxic acetaldehyde; on this basis one might predict that *aldA* should be only indirectly carbon-catabolite repressed via *alcR* and should have a basal level set higher than that of *alcA*. However, Northern analysis of strains with constitutive *alcR* expression (G6PD control) shows that both *alcA* and *aldA* mRNAs are directly glucose repressible (B. Felenbok, personal communication); hence *creA* acts at two levels to repress both *alcA* and *aldA* (Figure 1).

Thus all the components of this control system are genetically and physically characterized. The value of this relatively detailed knowledge is inestimable when considering the response of the system to the physiological changes associated with scaling up processes and changing medium and fermentation conditions. In contrast, the control gene(s) of the glucoamylase system have not been identified because of the total lack of genetic accessibility of *A. niger*.

III. ESTABLISHMENT OF THE *A. NIDULANS* SYSTEM

A. Initial Conditions

The choice of *A. nidulans* as the host species in which to undertake the primary development of an industrially competitive expression-secretion system was dictated by the high level of scientific development and genetic accessibility in that species. The key components of any expression system are the *cis*-acting and *trans*-acting control elements that determine its functional level and its response to the environment. Ideally, an expression system should be completely turned off when not in use, in order to avoid the toxic effects of foreign gene products and yet be capable of rapid induction by cheap and simple means to achieve high levels of expression. The *alcA* promoter has one of the lowest basal levels yet found for an inducible system, particularly when carbon-catabolite repressed, yet maximal gene expression is reached within 1 hr of induction by a shift from acetate to ethanol as the carbon source (Waring et al., 1989). To use a noncommercial example, each copy of an *alcA* promoter linked to a gene coding for β-tubulin (*tubC* or *benA*) directed the synthesis of 1% of newly synthesized cellular protein (Waring et al., 1989), measured after 8 hr of growth, and up to seven copies gave additive amounts of protein in these particular experiments. Thus the *alcA* and *aldA* promoters in the presence of the *alcR* gene product naturally have all the features of the ideal expression system. That this was likely to be the case was apparent based on the preliminary evidence on gene expression in various fungal control systems that was available in 1983.

Reasonably efficient transformation protocols and vectors carrying selectable markers were already available for *A. nidulans* from our laboratory (Tilburn et al., 1983) and others (Ballance et al., 1983; Yelton et al., 1984), 500 transformants or more per μg being easily obtained with most host strain-vector combinations. We used primarily the *argB* gene of *A. nidulans* as a selectable marker in both *A. niger* and *A. nidulans* and the *pyrF* gene of *A. nidulans* (and the *A. niger argB* gene once it had been cloned [Buxton et al., 1987]) for *A. niger* transformations. In general, cotransformation of a marker vector and a construction vector was used; generally, 80% of transformants for the marker gene are also transformed with the other vector. Filamentous fungi do not have nonchromosomal replicons with high copy numbers, such as are commonly used in bacteria and yeast. However, Dr. C. Scazzocchio and I had established (Tilburn et al., 1983) that transformants in which multiple copies of a vector were integrated into the host chromosome were easy to obtain, and simple plate tests showed that such multicopy transformants were stable through hundreds of mitotic divisions (F.P. Buxton, personal communication), i.e., certainly stable enough to maintain copy number through a fermentation run. Ours and many subsequent studies in various laboratories have now shown that a considerable number of the copies present in multicopy transformants occur as arrays of vector molecules in tandem at a given chromosomal site, with a background of single or low-multiple integration events elsewhere. If the vector contains sequences homologous to chromosomal sequences, then from 10% to 40% of the integration events will be at the site of homology, depending on its extent. Transformants with 10-20 copies of the vector are common, and strains with around 50 integrated copies are easy to obtain by screening. Higher numbers are occasionally obtained by more extensive screening or by passing tandem arrays through meiosis (meiosis destabilizes multicopy transformants; copy number is mostly reduced, but occasional colonies show increases). All multicopy transformants used in this work were obtained by direct screening of transformant DNA in Southern transfers containing a single-copy internal standard.

Concern is often expressed about the potential danger of the extracellular proteases of *Aspergillus* to the recovery of protein products. In fact, all the secreted proteases are very strongly nitrogen-metabolite regulated, so that in the presence of sufficient levels of ammonium ions (threefold above the standard medium concentration), the level of secretion of these proteases is very low. In addition, a strain carrying a nonreverting mutation in the *areA* gene, which codes for the positively acting regulatory protein essential for the expression of these (and many other nonessential) genes, can be used to prevent protease expression; such strains grow well. Degradation by proteases has not proved to be a problem in practice so far, so that contingency plans that were made to deal with the potential problem were never activated.

B. Strategy

The overall strategy was to develop and test the expression system in *A. nidulans* using the *alcA-alcR* system, while developing the tools for switching to *A. niger* as a host should this prove desirable in terms of yield or for commercial reasons. It was decided to develop the use of the *A. niger* glucoamylase promoter in parallel, firstly because this promoter (or variants there of) is known to be capable of industrial levels of enzyme production and secondly because it is an *A. niger* promoter, and at the time the degree of relatedness of *A. niger* and *A. nidulans* was unclear. In the absence, at that time, of direct information about secretion signal sequences in *A. nidulans*, it was decided to use a generic optimized signal sequence for the secretion vector constructions, while cloning genes from secreted proteins of *A. nidulans* or *A. niger* to provide homologous signal sequences. The phosphatases were chosen as a target because of the genetic information available, and because genes such as *pacA* (acid phosphatase) are regulated not only by inorganic phosphate concentrations, but also by pH in both *A. nidulans* (Caddick et al., 1986) and *A. niger*, and this might form the basis of a feasible large-scale industrial process. In fact, both the artificial generic signal sequence and the *A. niger* glucoamylase signal sequence function very well in both *A. nidulans* and *A. niger*.

IV. CONSTRUCTION AND TESTING OF EXPRESSION-SECRETION VECTORS USING THE *alcA* PROMOTER

A. First-Stage Vectors

Two series of secretion vector constructions were carried out (Gwynne et al., 1987b), one based on the *alcA-alcR* control system of *A. nidulans* and the other on the glucoamylase control system of *A. niger*. The foreign genes used in developing the system were a gene from the Gram-positive bacterium *Cellulomonas firmi* encoding a secreted endoglucanase, which had been shown in our laboratory (Skipper et al., 1985; 1987) to be well secreted from yeast, and the human interferon α2 gene.

The complete sequence of the *alcA* gene and a considerable amount of flanking region was determined (Gwynne et al., 1987a), and deletion analysis showed that all the *cis*-acting sequences necessary for *alcR*-mediated induction of the *alcA* promoter lay within 320 bp of the translation start. The initial expression vector pALCA1 and the expression-secretion vector pALCA1S are based on the *E. coli* vector pUC12, and actually contain 2 kb of the *alcA* sequence upstream of the translation start, except for the last 50 nucleotides of the leader region (the transcription initiation sites are of course present), which was removed by exonuclease III resection. pALCA1 has *Sal*I and *Xba*I cloning sites immediately downstream of the *alcA* promoter and the transcription start. Instead of the *alcA* leader and the N-terminal coding region, a sequence coding for an artificial signal sequence was incorporated from pALCA1S (Gwynne et al., 1987b) (Figure 4). The vector pALCA1SL has a polylinker sequence immediately downstream of the secretion signal. The synthetic secretion signal used was designed according to Von Heijne's rules (Von Heijne, 1986), with two amino-terminal postively changed residues, a hydrophobic core, and an optimized cleavage-site sequence. No transcriptional terminator was incorporated into these prototype vectors.

The human interferon α2 gene was taken as a *Dde*I-*Bam*HI fragment from pN5H8 (Slocombe et al., 1982), which leaves a region coding for a fragment of the interferon signal peptide at the 5′ end of the interferon gene. The in-frame fusion of this interferon sequence to the synthetic signal peptide in the resulting vector pALCA1SIFN (Gwynne et al., 1987b) thus codes for 16 amino acids between the last amino acid of the synthetic signal and the first amino acid of mature interferon 2, nine of which derive from the interferon signal peptide and seven from sequence accumulated during vector constructions. Nevertheless, cotransformation of *argB-pyrG*89 *A. nidulans* with pALCAIS1FN and pDG2 (Buxton et al., 1985), which carries *argB⁺*, yielded a high proportion of transformants that expressed human interferon α2 on induction of the *alcA* promoter with threonine. In the particular experiment that was published (Gwynne et al., 1987b), 17 of 20 *argB⁺* transformants contained sequences homologous to a human interferon DNA probe, varying from one to approximately 20 copies per genome (Figure 5 shows equivalent results with the glucoamylase

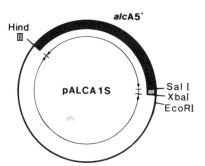

Figure 4 Restriction map and composition of the expression-secretion plasmid pALCAIS. The hatched area represents DNA encoding a synthetic secretion signal sequence. The construction of this plasmid is described in detail by Gwynne et al. (1987b), and subsequent modifications of it are described in that reference and in this text.

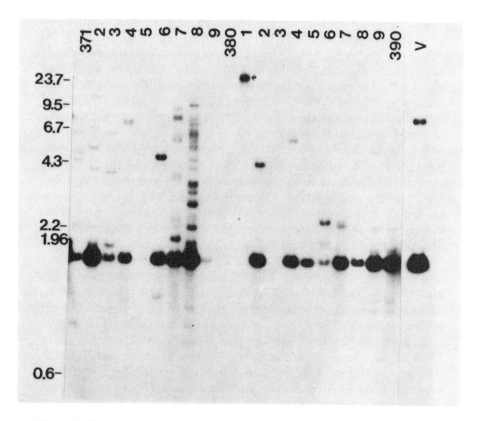

Figure 5 Southern transfer of DNA from *Aspergillus nidulans* transformants for the human interferon α2 gene under the control of the *A. niger* glucoamylase promoter. Genomic DNA was isolated from 20 well-separated *A. nidulans* colonies, which had been cotransformed to prototrophy with *argB* along with pGL2BIFN (Gwynne et al., 1987b). Equal quantities (10 μg) of DNA were cut with the restriction endonucleases *Bam*HI and *Sst*I, electrophoresed on an 0.8% TAE agarose gel, and transferred to Genescreen Plus. The blot was probed with a *Bss*HII-*Sst*I fragment from plasmid pGL2BIFN containing the interferon insert. Twenty *argB*⁺ transformants are indicated (lanes 371-390). Vector pGL2BIFN was cut with the restriction endonucleases *Bss*HI and *Sst*I and used as a reference (lane V). Lambda DNA fragments from a *Hin*dIII digestion were used as size markers, and these are indicated on the left in kilobase pairs.

promoter vector). As with all such experiments in *A. nidulans*, the majority of the multiple copies of pALCA1SIFN occurred as tandem arrays. Eleven of the 17 transformants containing interferon $\alpha2$ DNA showed detectable levels of secreted interferon $\alpha2$ by immunoassay, when the transformant strains were grown on minimal medium containing 1% fructose (a nonrepressing carbon source) and were induced with 100 mM threonine in 0.1% fructose. Interferon synthesis was also shown to be subject to carbon-catabolite repression (Figure 6), although there were variations in the relative levels of expression on glucose, fructose, and fructose plus inducer compared to typical data for the *alcA* gene itself. This is not surprising in view of the differences in chromosomal location of integrated interferon genes and the occurrence of a background of rearranged sequences that could come under different control. However, this experiment, and others subsequently, showed that it is quite satisfactory to proceed empirically and screen a relatively low (20-50) number of transformants in order to find one that has the desired phenotype.

The same experiment, with very similar results, was carried out with the *Cellulomonas fimi* endoglucanase gene (actually a fragment lacking the NH_2-terminal coding region; Skipper et al., 1985) fused to the *alcA* promoter and synthetic signal. Once again the majority of transformants produced measurable levels of extracellular bacterial endoglucanase, in an *alcR* and inducer-dependent manner subject to carbon-catabolite repression. Secretion was also shown to be dependent on the synthetic secretion signal, since transformants with a different vector, pDG6, had the endoglucanase gene C-terminal coding fragment fused to a fragment of the *alcA* gene coding for the NH_2-terminal 80 amino acids of ADHI, with no secretion signal, and endoglucanase activity was only detectable intracellularly (Gwynne et al., 1987b).

Thus directed, completely controllable secretion of two foreign proteins from *A. nidulans* (and *A. niger*, see below) was achieved. Moreover, the yield of secreted endoglucanase was quite reasonable for growth in minimal medium. It was estimated at 1 mg/l from cultures grown to about 0.5 g/1 dry weight in fructose-threonine medium, based on all the secreted protein being active and possessing the same specific activity as the bacterial product. The level of interferon $\alpha2$ was 20-fold lower, but this was in part due to the construction. Most of the secreted interferon $\alpha2$ was biologically active, as assessed by the plaque reduction assay (Fish et al., 1983), provided again that the specific activities of fungal and human cell products are equivalent.

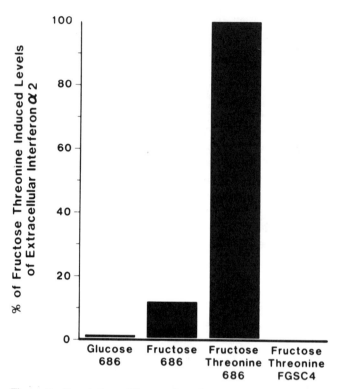

Figure 6 Regulation of human interferon α2 expression and secretion in transformants containing this gene under *alcA* promoter control. *Aspergillus nidulans* strain 1000, which had been contransformed to prototrophy with *argB* along with plasmid pALCA1SIFN, and was selected as expressing extremely high levels of IFN α2 compared to other transformants, was grown along with the wild-type strain (FGSC4) at 37 °C on minimal medium plus 1% glucose, 1% fructose, or 0.1% fructose and 100 mm threonine. Samples of growth medium were removed at 48 hr and assayed for protein levels and interferon activity. Interon activity levels are expressed as the percent of maximally induced specific activity levels of strain 1000 on fructose threonine medium.

B. Second-Stage Vectors

A number of simple improvements were made to the prototype *alcA* promoter vector pALCA1S and its derivatives.

1. The transcriptional terminator of the *A. niger* glucoamylase gene was included in later vectors. In fact, interferon α2 and other mRNAs in pALCA1S showed a fortuitous major termination site about 200 nucleotides

into vector sequences downstream; thus it was not too surprising that vectors including the glucoamylase transcription stop did not give higher yields.

2. All constructions are made as precise fusions of the signal peptide sequence with the coding sequence of the foreign gene at the desired amino acid. In a few cases (particularly totally synthetic genes), this was facilitated by introducing particular restriction sites at the end of the signal sequence coding region in place of the original polylinker, e.g., the vector pTAWTS (Gwynne et al., 1989), a derivative of pALCA1S that has a *Dra*I (or *Nae*I for blunt-end cloning) site at the 3′ end of the secretion-signal sequence, allowing direct fusion between the end of the signal and the desired mature polypeptide if the DNA fragment has an *Fsp*I (or any blunt-end) site at the 5′ end. In another case, the expression-secretion vector pALCA1SL was modified by in-vitro mutagenesis so that the ATG is contained within the sequence of an *Nco*I site and the secretion signal is deleted: the resulting expression (nonsecretion) vector pALCA1L*Nco*I allows the generation of an ATG start codon when a DNA fragment with a 5′ *Nco*I end is cloned into the *Nco*I site. In general, however, many constructions were made by taking an approximate initial fusion with some material between the signal peptide coding sequence and the coding sequence of the foreign gene, and introducing a deletion via oligonucleotide-directed in-vitro mutagenesis (Kunkel, 1985) to produce a precise fusion. The value of this is clearly shown by the fact that when the 16 amino-acid stretch between the signal peptide and the beginning of mature interferon α2 in the protein produced from pALCA1SIFN was removed by precise deletion at the DNA level, the yield of secreted interferon α2 increased 15-fold (Gwynne et al., 1987b). Similarly, quantitative comparison of the amounts of interferon α2 secreted per integrated copy of a precise-fusion acid phosphatase promoter-inferferon α2 construction and an imprecise acid phosphatase-interferon construction (with two amino acids from mature acid phosphatase, nine from the interferon signal, and two from a linker between the signal peptide and the interferon gene) also showed a clear difference: the precise fusion produced on average 50 times more secreted interferon in *A. nidulans* and threefold more in *A. niger* (MacRae et al., 1989).

3. All current *alcA* promoter vectors have the entire natural *alcA* mRNA leader sequence and the natural sequence prior to the ATG. The production of these vectors from pALCA1S took place in two stages. A synthetic DNA fragment was inserted into the *Sal*I site that is immediately upstream of the ATG in pALCA1S, restoring the sequence coding for the *alcA* leader, except for eight bases before the ATG; this step produced no

yield increase of interferon $\alpha 2$. However, restoring the *alcA* sequence just before the ATG by replacement in vitro mutagenesis produced an average twofold yield increase in 20 transformants for interferon $\alpha 2$. The *alcA* sequence contains the A at -3, which is characteristic of efficiently translated genes (Kozak, 1984, 1986).

4. The size of the 5′ flanking region of the *alcA* gene was reduced to 320 bp in one set of vectors. This region of the *alcA* promoter contains all *cis*-acting elements that define *alcR* activation and transcription initiation, but not a putative *creA* binding site (M. Devchand, personal communication). The basal levels of transcription and the parameters of induction have not yet been rigorously determined for this version of the *alcA* promoter, but are close to the wild type.

V. THE *A. NIDULANS alcA* EXPRESSION-SECRETION SYSTEM AND ITS USE

The *A. nidulans* expression-secretion system currently in use has three components: the set of second-stage vectors described above, a set of host strains for expression, and a set of fermentation conditions.

When standard *A. nidulans* strains were transformed with vectors expressing interferon $\alpha 2$ under control of the *alcA* promoter, the yield of interferon $\alpha 2$ increased in a highly variable, but basically linear, fashion up to about 7-10 integrated copies of the vector, but not thereafter. In addition, the level of transcription from the intrinsic *alcA* gene on induction was reduced, as was the level of ADHI in multicopy *alcA* promoter-IFN $\alpha 2$ transformants; such strains were relatively resistant to allyl alcohol, which is converted to the toxic product acrolein by ADHI. A likely interpretation of these results is that titration of the *alcR* product by multiple *alcA* promoters led to a reduction in expression in the intrinsic *alcA* gene and to limitation of the expression from multiple *alcA* promoters linked to interferon $\alpha 2$. In order to provide further data, the behavior of transformants with a range of copy numbers of the *alcA* promoter driving the *alcA* gene itself, in the presence of different numbers of copies of the *alcR* gene, was investigated in a very pragmatic way (F.P. Buxton and R.W. Davies, unpublished); i.e., the copy number of the *alcA* and *alcR* genes was determined only at the DNA level by comparison with the intensity of the wild-type single-copy band in Southern transfers, and ADHI enzyme activity was the parameter measured. Although the results (Table 2), and their detailed interpretation are complex, the following conclusions could be drawn: the presence of multiple copies of *alcA* in strains with only one copy of *alcR* had little effect on ADHI levels; increasing the copy number

Table 2 Illustration of the Effect of Copy Number of *alcA* and *alcR* on ADHI Expression in *A. nidulans*

Strain	Copy number[a]		Specific activity of ADHI[b]
	alcR	*alcA*	
FGSC4[c]	1	1	10
1[d] (T580)	50	1	20
2 (T632)	1	50	15
3	a	e	10
4	a	e	70
5	a	e	19
6	b	e	100
7	c	e	100
8	c	e	145
9	d	e	130
10	d	e	140
11	d	e	60
12	d	e	140
13	e	c	100
14	e	c	100
15	e	c	230[e]
16	e	a	30
17	e	a	60
18	e	a	20
19	e	a	90
20	e	a	90
21	e	a	205[e]

[a]Copy number was estimated approximately by comparison of band intensities on Southern transfers of transformant DNA to single-copy, wild-type DNA. These data are given for the illustration of general trends only. Where a number is not given, a to e represent levels of increasing copy number: a, < 5; b, 4-10; c, 10-20; d, 20-40; and e, ~ 50.

[b]Specific activity of ADHI was measured after induction as described by Lockington et al. (1985).

[c]FGSC4 is a wild-type strain of *A. nidulans*.

[d]T580 (multiple *alcR*, single *alcA*) was transformed with pFB39 (*argB*) and a plasmid carrying the entire *alcA* gene and flanking regions to generate transformants 13-21. T632 (multiple *alcA*, single *alcR*) was similarly cotransformed with *argB* in pFB39 and a plasmid carrying the entire *alcR* gene and flanking regions, generating transformants 3-12.

[e]Note that transformants 15 and 21 are exceptional: transformant 15 conforms to the general trend, but has by far the highest level of expression in the series; transformant 21 has relatively few copies of *alcA*, but produces almost as much ADHI as 15. It is obvious that the *alcA* and *alcR* copy numbers are not the only determining factors for ADHI expression.

of *alcR* had no effect on the expression of a single *alcA* gene; in strains with a high copy number of *alcA*, increasing numbers of *alcR* genes lead to increasing levels of ADH; and in strains with a high copy number of *alcR*, increasing copy numbers of *alcA* lead to increasing levels of ADHI. All strains with multiple *alcA* and multiple *alcR* genes were more sensitive to allyl alcohol than was the wild type, grew less well on ethanol, and failed to conidiate on this carbon source, presumably because of the accumulation of acetaldehyde. Certain multiple *alcR*, multiple *alcA* strains produced over 20 times the wild-type level of ADHI after induction. These experiments are being repeated in a rigorous fashion. Strains carrying multiple copies of pALCA1SIFN 2 were also converted to multiple *alcR*; in such strains the level of secreted IFN α2 continued to be approximately linearly related to the copy number of pALCA1SIFN 2 up to at least 20 copies (Gwynne et al., 1987c). Thus, although the variation between isolates is very great and strains have been found that appear to be exceptions to the general trend, all development work from this point onwards was undertaken in host strains that are *argB pyr89* multiple-*alcR* derivatives of FGSC4, e.g., T580, a multi-*alcR* single *alcA* strain with about 50 copies of *alcR* (Table 2). T580 has essentially normal levels of ADHI and has been shown to produce more *alcR* gene transcript than does the wild type. In subsequent experiments involving the expression of other foreign genes under the control of the *alcA* promoter, considerably higher levels of transcription from integrated copies of the foreign gene were consistently observed in T580 compared to the wild type as the host.

It is not the purpose of this article to discuss fermentation procedures, but these are of considerable importance in determining the final performance of an expression system. Since the system as it stands is subject to strong glucose repression, protocols now in use separate the induction phase from the initial rapid accumulation of biomass on complex media containing glucose or other repressing carbon sources. Thus, towards the end of the first phase of biomass accumulation, glucose levels are depleted sufficiently to allow induction. Inducer is then added together with a non-repressing carbon source. Initial testing of fermentation conditions was undertaken using secretion of *A. niger* glucoamylase driven by the expression of the *alcA* promoter in a multi-*alcR* strain of *A. nidulans*, since it is reasonable to assume that glucoamylase is well adapted for secretion from fungi, allowing one to concentrate on the control of gene expression as a major variable (Gwynne et al., 1987c). In one experiment, for example, 14 gm/l of biomass was accumulated in the 30 hr that elapsed before glucose was sufficiently depleted to allow induction and after a 10- to 15-hr postinduction lag phase, secreted active glucoamylase accumulated to 1.2 g/l (assuming that the specific activity of the *A. nidulans* product is the same as the *A. niger* product). This is quite a respectable amount of protein. Nevertheless, it is clear that very many variables can be altered in such experiments, from which inducer to use and the levels of trace elements, carbons, and nitrogen sources in each phase, to aeration levels and fermentor design. In general, the fermentation development of each protein product has to be viewed as a separate problem, with common denominators emerging with experience.

A number of foreign proteins have been produced in *A. nidulans*, the majority from the *alcA* promoter (Table 3). The details of most of this work is unfortunately, to some degree, confidential. In a typical series of experiments, a synthetic epidermal growth factor (EGF) gene was fused precisely to the signal sequence of the 5′ corrected version of pALCA1S (Section IV.B), and multicopy pALCA1SEGF transformants of multi-*alcR* strain T580 were found by Southern screening. In minimal medium the best strains secrete 1-2 mg/l of correctly processed human EGF, at about 0.5 g dry weight of mycelium/l. Experience with scaling up the production of other proteins suggests that it is reasonable to expect to obtain at least 100 mg/l in fermenter runs with such strains after conditions have been optimized. The EGF product was active in receptor binding assays and was detectable by radioimmunoassay, implying that the product is folded correctly. Since mature EGF contains three disulfide bonds that are essential for correct conformational folding and biological activity, the *A. nidulans* system would seem to provide a valuable alternative to *E. coli*, where the inefficiency of formation of disulfide bonds in foreign proteins is a well-documented phenomenon.

Table 3 Heterologous Proteins Expressed in *Aspergillus nidulans*

Species of origin	Protein	Secretion
Bovine	[b] Prochymosin	+
Human	[c] Tissue plasminogen activator (TPA)	+
Human	[a] Superoxide dismutase	−
Human	[a] Epidermal growth factor (EGF)	+
Human	[a] Parathyroid hormone (PTH)	+
Human	[a] IL6	+
Human	[a] Growth hormone	+
Human	[a] Corticosteroid-binding globulin (CGB)	+
Human	[a] Interferon 2	+
Bacterial (Cellulomonas fimi)	[a] Endoglucanase	+

[a]Expressed from the *alcA* promoter.
[b]Expressed from the *A. niger* glucoamylase promoter.
[c]Expressed from the *A. niger adhA* promoter, and *A. nidulans alcC* and *tpiA*.

Although this system was designed for the secretion of foreign proteins as the basis of the production regime, it has proved to be equally useful for intracellular production of human proteins. *AlcA* promoter vector lacking the signal peptide, such as pAlCA1 or pALCA1L-*NcoI* (see above), can be used for intracellular production. Initial transformants containing multiple copies of these vectors expressing a gene for the nonsecreted human enzyme with several established medical uses directed the intracellular accumulation of 3-5 mg/l of this protein on growth and induction in minimal medium, and it proved possible to produce 400 mg/l of this protein from this strain after preliminary optimization of fermentation conditions. There seems to be no reason why this protein should not soon be produced at a level comparable to that of homologous *Aspergillus* proteins that have industrial uses.

VI. *ASPERGILLUS NIGER* EXPRESSION AND SECRETION SYSTEMS

While the *A. nidulans* system was clearly the most amenable to rational development using molecular biology and genetics, *A. niger* is of such great industrial importance that the development of an *A. niger* expression-secretion system had to be a parallel goal. In contrast to the strategy adopted for *A. nidulans*, where a well-characterized standard laboratory strain was used as the basis for system development, it was decided that in the case of *A. niger* we should just concentrate on developing tools that allowed any preexisting strain of *A. niger* to be used as an expression host, and then use these tools to transfer the components of the *A. nidulans* system into *A. niger* if this proved to be superior for certain purposes.

A. Transformation Systems: *argB* and *sC*

When this program was initiated, transformation of *A. niger* had not been reported. We decided to rely upon *A. nidulans* and *A. niger* being sufficiently closely related for genes of one species to be expressed to some extent in the other. The key components of a useful transformation system are a cloned selectable gene and a recipient in which expression of the marker can be selected. Dominant selectable markers such as hygromycin resistance are now available in *Aspergillus*, but in their absence at that time we developed a recipient strain carrying a complementable mutation with a low reversion frequency. Of the *A. nidulans* genes available for complementation, we chose to work with the *argB* gene (Berse et al., 1983), which codes for ornithine carbamoyl transferase and was known to produce an active product in yeast. Defects in amino acid biosynthesis produce very clear phenotypes, so that even poor complementation with *argB* is easy to detect. *A. niger argB* mutants were generated, and transformation of *A. niger argB* was successfully achieved (Buxton et al., 1985) with plasmids carrying *argB* of *A. nidulans*, using protoplasting and transformation conditions optimized for *A. nidulans*. This was the first successful transformation of this organism, together with the simultaneous analogous work using *amdS* as the complementing marker (Kelly and Hynes, 1985). We rejected *amdS* as a marker, since some *A. niger* strains utilize acetamide too well to allow the selection of transformants.

A. *niger* transformants behave exactly like the better-characterized *A. nidulans* transformants, i.e., all stable transformation is by integration, integration is not necessarily dependent upon extensive homology, integrative transformants are stable through many mitotic generations, a variety of integration sites is used, and multiple tandem integration is common. However, detailed investigation has not been undertaken. The frequency obtained has never been better than 10-20 transformants per microgram, which is at least an order of magnitude below the *A. nidulans* frequency. This may reflect a need for optimization of protoplasting and protoplast regeneration conditions. Another possibility, which was raised by the relatively high frequency of occurrence of small-colony unstable transformants (usually attributed to temporary persistence of nonintegrated vector), is that the mitotic recombination level in *A. niger* is lower than that of *A. nidulans* and/or more dependent on homology. We subsequently cloned the *argB* gene of *A. niger* by complementation of *argB* of *A. nidulans* (Buxton et al., 1987) and constructed a transformation vector using the *A. niger* gene (F.P. Buxton and R.W. Davies, unpublished). Neither this vector, nor vectors in which random fragments of *A. niger* DNA of various lengths were incorporated, gave a significant increase in the transformation frequency of *A. niger*, indicating that the lack of homology is not the problem (F.P. Buxton and R.W. Davies, unpublished). It should be pointed out that this transformation frequency, while low for purposes such as selection of clones from gene libraries (which can be performed in *A. nidulans*), is quite adequate for the introduction of any defined genetic construct into the genome of *A. niger*.

The real goal of transformation system development in *A. niger* is to become able to deliver genetically engineered constructs to industrial strains in order to take advantage of the special properties these strains have gained through long-term mutagenesis and selection. Generating auxotrophic markers is tedious, and some strains are known to be polyploid, making this approach very difficult, if not impossible. We have developed the use of selection for and against ATP sulfurylase function to solve the problem of how to make all strains of *A. niger* accessible to genetic transformation (Buxton et al., 1989). We showed that mutants of *A. niger* resistant to selenate are easily obtained. As in *A. nidulans*, these mutants carry defects in one of two genes: *sB*, deficient in sulfate permease, selenate and chromate resistant; or *sC*, deficient in ATP sulfurylase, selenate resistant, but chromate sensitive. Thus *sC* mutations in any *A. niger* strain can be obtained by using the very powerful selection for selenate resistance on large numbers of vegetative spores mutagenized simply by UV irradiation, followed by testing for chromate sensitivity. In a typical experiment with UV irradiation to 90% kill (Buxton et al., 1985), plating 3×10^5 spores gave rise to 23 selenate-resistant colonies, of which 21 were sC^- and two sB^-. These sC^- *A. niger* strains are then used as transformation recipients for vectors carrying the *sC* gene of *A. nidulans*. Since sC^- *A. niger* do not utilize sulfate, they will not grow on media containing sulfate as the sole source of sulfur, whereas transformants containing the *A. nidulans* *sC* gene will. Cotransformation, which works perfectly well in *A. niger* despite the low frequency of transformation [one integrative transformant per microgram of vector with the *A. niger* strains used (Buxton et al., 1989)], allows one to deliver any DNA to any *A. niger* strain in this way. The *sC* vector is pFB95; it carries the *sC* gene of *A. nidulans*, which is expressed in *A. niger*.

B. Expression Systems

Two expression systems were developed that are based on inducible genes of *A. niger* that code for secreted proteins. Glucoamylase is itself an industrial enzyme, and strains of *A. niger* exist that produce this enzyme at 5 g/l (and, according to hearsay, at even higher levels) under appropriate conditions. The choice of the *pacA* gene, coding for acid phosphatase, was based on the good level of genetic analysis of the control of this gene in *A. nidulans* and the potential convenience of its inducibility by pH in fermentation development.

1. Glucoamylase

The glucoamylase gene, which was have named *glaA*, is both glucose repressible and inducible in the presence of starch (by an unknown metabolite) to very high levels. The gene was cloned (Gwynne et al., 1987b) using redundant oligonucleotide probes created by reverse translating the first six amino acids of the published amino-terminal sequence of the mature glucoamylase protein. By identifying the DNA sequence corresponding to the mature amino terminus, the signal peptide and 5′ flanking sequences were located. The occurrence of a *Bss*HII site coincident with the site of signal peptide cleavage allowed the insertion of a set of linkers that provided for fusion of any foreign coding sequence to the signal peptide in place of the glucoamylase sequence. These vectors (pGL2A, B, and C) are still useful for initial experiments, although we now prefer to produce a precise fusion to the signal peptide. Remarkably, the glucoamylase promoter proved to be glucose repressible and starch-metabolite inducible in *A. nidulans* when tested. This was useful, as it provides us with an inducible strong promoter as an alternative to the *alcA* promoter. Since *A. niger* and *A. nidulans* seem, through our experience, to be remarkably similar in the details of their fundamental molecular biology, the conservation of the recognition signals for a control system like glucose repression that affects many genes is comprehensible; the maintenance of the recognition signals for starch control is rather more surprising. Using this system, the two test proteins (*C. fimi* endoglucanase and human interferon α2) were produced in active extracellular form from *A. nidulans*. Plate tests of transformants of *A. niger* with derivatives of these vectors expressing the *C. fimi* endoglucanase showed that the glucoamylase promoter maintained its normal control behavior when linked to a foreign gene. Quantitative analysis was carried out in *A. nidulans* and showed that the levels of secreted protein generated by the *alcA* and *A. niger glaA* promoters are very similar.

2. Acid Phosphatase (*pacA*)

Acid phosphatase is a secreted enzyme of both *A. niger* and *A. nidulans*, the synthesis of which is regulated by the concentration of inorganic phosphate and by the pH of the culture medium (Caddick et al., 1987a, 1987b). These enzymes are convenient to work with genetically, since simple staining methods allow the detection of the enzyme both in mycelia and after secretion into the surrounding (solid) medium. It has been shown (Caddick and Arst, 1987) that *pacA* is the structural gene for an intra- and extracellular phosphate-repressible phosphomonoesterase, and that phosphate repression control of *pacA* and other genes is mediated by the positive-acting regulatory gene *palcA*. Under phosphate-derepressed conditions, the expression of *pacA* is enhanced at low pH and decreased at high pH. An *A. niger* gene coding for an acid phosphatase was cloned by complementation of *A. nidulans pacA*101 by a plasmid in an *A. niger* gene library (MacRae et al., 1988). The transcription of this gene is regulated by the inorganic phosphate concentration in the growth medium when the gene is integrated into the *A. nidulans* genome.

Thus, as with *glaA*, a control system present in *A. nidulans* (presumably *palcA*) retained specificity for the recognition sites of an *A. niger* gene under the same control system in another species. The gene cloned has been named *pacA* of *A. niger*. A 3.05kb *Xba*I-*Bam*HI fragment containing this gene was sequenced and the transcript mapped, allowing us to locate the promoter region and the coding sequence, including the signal sequence. The maintenance of regulation of *pacA* in transformants with this fragment indicated that the 416-bp region upstream from the transcriptional start point to the *Xba*I site contains the necessary signals for initiation of transcription and phosphate regulation. The signal peptide cleavage point was assigned based on the weight-matrix method of Von Heijne (Von Heijne, 1986) and comparison to other fungal cleavage sites, and secretion vectors were constructed on the expectation that Ala18/Gln19 is the site of cleavage. The upstream noncoding region to the *Xba*I site, and the signal-peptide coding region extending to the Gln19 codon was fused precisely to the human interferon α2 gene by initial cloning followed by deletion in-vitro mutagenesis. *ArgB* was cloned in, and the vector pACIF2 was used to transform, *argB A. niger* and *A. nidulans*. Transformants of both species were easily obtained that secreted interferon into the culture medium. Interferon mRNA synthesis and interferon secretion were controlled by the concentration of inorganic phosphate in the growth medium. The induction ratios were 5.6 and 6.2, respectively, for interferon mRNA in the *A. nidulans* and *A. niger* transformants that were studied in detail, with the induced level of interferon in mycelium being 5- and 18-fold higher. Regulation of interferon expression by pH, with significantly greater (5 x) levels of interferon synthesis at pH 4.5, relative to pH 8.5, was demonstrable for the *A. niger* transformant. It is interesting that the *A. nidulans* pACIF2 transformant secreted 87% of the interferon it produced and 42% of the intrinsic acid phosphatase, whereas the *A. niger* transformant only secreted 11% and 6%, respectively. While it seems unlikely that all *A. niger* isolates are such inefficient secretors, this result does reinforce the correctness of the decision to use *A. nidulans* for production. Thus the *A. niger* acid phosphatase gene promoter and signal sequence can be used to direct the synthesis and secretion of foreign proteins under environmental control by phosphate concentration and pH. The level of mRNA synthesis directed by this promoter is, unfortunately, low compared to the *alcA* promoter. The induced acid phosphatase promoter accounted for 0.01% of the total RNA of *A. niger*, and the same level was produced per copy by the *A. niger* and *A. nidulans* transformants studied. This obviously represents a higher proportion of newly synthesized mRNA, and transformants have been found with 10-fold higher levels of mRNA synthesis from the induced *pacA* promoter per copy. Nevertheless, this system has not become the subject of further industrial process development work.

3. *alcA*

In all the experiments where *A. niger* and *A. nidulans* transformants have been compared, *A. nidulans* has performed as well or better than *A. niger*. Since there was a shift in commercial emphasis to producing therapeutic proteins rather than enzymes or other food industry products, no particular reason has arisen to emphasize *A. niger* as a protein expression host, contrary to initial expectations. Therefore the planned transfer of the *alcA-alcR* system into *A. niger* has not been undertaken. Transformation of *A. niger* with constructs containing the *alcA* promoter showed that the *alcA* promoter does not function at significant levels in *A. niger*, presumably due to the lack of *alcR* product. The obvious experiment of supertransforming *A. niger* containing the *alcA* promoter linked to a reporter gene, with the *alcR* coding region linked to a promoter known to be active in *A. niger* (such as that from the *A. nidulans* glucose-6-phosphate dehydrogenase gene or the *A. niger* glucoamylase promoter), has not been done. If needed, this should certainly make the *alcA* promoter system usable in *A. niger*.

VII. MOLECULAR BIOLOGY AND FUTURE DEVELOPMENT OF THE *ASPERGILLUS* *alcA/alcR* EXPRESSION SYSTEM

It is clear that the ongoing development of any expression system must involve the interplay of molecular biology, genetics, microbial physiology, and fermentation practice. The system(s) described here, while designed according to the best available information, are essentially assemblages of natural elements that in combination confer desired properties on the system. To paraphrase the words of Sydney Brenner, they are the work of genetic mechanics, not genetic engineers. Since the *A. nidulans alcA* system performed well enough to raise expectations that respectable levels of production of certain proteins could be achieved by optimizing the growth conditions and fermentation procedures, emphasis has naturally been placed on using the available system commercially. The system, therefore, relies on the natural good performance of the *alcA* promoter and of *A. nidulans* as a host for gene expression and protein secretion. The real future of the system lies not only in wearing out the seats of fermentation engineers' and microbial physiologists' pants, but also in going beyond the natural capacity of the system by true genetic engineering.

From the beginning of this program, it was clear that four areas of basic knowledge would be particularly relevant to the further improvement of the system: transcription and its control, mRNA stability and translation, secretion, and interrelationships between the elements and products of the expression system and the environment inside and outside the host. The Allelix group put a small but continuous effort into increasing our understanding of some of the above areas of fungal molecular biology. The progress to date will be briefly described, since much of it is relevant to all fungal expression and secretion systems, and some indication of future directions will be given.

A. Experimental Approaches

1. Promoter Analysis

The *alcA* (and *aldA*) promoters are obviously key elements in the system. We have investigated the functional elements within these promoters, with two major goals: short term, the unraveling of *alcR* and *creA* (carbon catabolite) control; long-term, improvement of the performance of these promoters. In order to ask precise questions about the role of particular sequence elements, it is desirable to construct a system that allows homologous replacement of the natural gene with alleles that have been altered in defined ways in vitro. A marker gene (with its own functional promoter) was inserted into the coding region of plasmid-borne *alcA* or *aldA* in vitro. The majority of the experiments used *pyr4* of *Neurospora crassa* as the disrupting marker gene, since it is possible to select both for and against the function of this gene. Transformation of a *pyr argB* recipient with disrupted-*alcA* plasmid linearized in *alcA* (or a disruption fragment excised from the plasmid by cuts in *alcA* or its linking regions) should yield transformants with a disrupted chromosomal copy of *alcA* (or *aldA*), after screening *pyr*+ transformants for *alcA*- and checking on Southern transfers. Once such a strain is obtained, modified versions of the *alcA aldA* gene can be inserted into the correct chromosomal locus by homologous recombination, replacing the marker gene by the *alcA(aldA)* sequence; in the approach used, the mutant *alcA* allele would be introduced on an *argB* plasmid by selection for *argB* and counterselection for *pyr*-. Unfortunately, the disruption strains were extremely difficult to construct; eventually *pyr4* was abandoned as a disruption marker and *argB* was used. With this disruption vector, a strain with the chromosomal *aldA* gene disrupted by *argB* was obtained; this will allow precise quantitative experiments on this promoter to be carried out (M. Saunders, unpublished results). It is possible that the difficulty experienced in obtaining homologous replacement of these genes is related to the extremely low levels of transcription of *alcA* and *aldA* when glucose is repressed and uninduced, since a significant correlation of homologous replacement frequency with transcriptional state has been found in mammalian systems (Capecchi, 1989). Meanwhile, qualitative experiments on these promoters were carried out without the precise replacement system. The sequencing of the *alcA* and *aldA* genes had been achieved using processive deletion strategies (Gwynne et al., 1987a; Pickett et al., 1987), so that a number of deletions with end points within the putative promoter region were available. In addition, particular sequences identified as occurring in roughly equivalent positions in the *alcA* and *aldA* promoters (Figure 2) were deleted precisely by oligonucleotide-directed in-vitro mutagenesis. A number of single-copy transformants were made for each mutant form of the *alcA* or *aldA* promoter; while the quantitative performance of such transformants is of little significance, the occurrence of consistent data across all transformants for *alcR* dependence and carbon-catabolite repression allows conclusions to be drawn. In one set of experiments, transformants were made in normal *alcR* hosts and scored for *alcA* or *aldA* gene expression phenotype (*alcR* inducibility, constitutivity, or lack of expression). Preliminary results (M. Devchand and M. Saunders, personal communication) indicate that sequence C (Figure 2) corresponds with the binding site for *alcR* protein. The *alcR* promoter and details of the binding of *alcR* protein to DNA are being studied by Drs. B. Felenbok, C. Scazzochio, and colleagues.

In another set of experiments, a host strain was used that was itself a transformant for an *alcR* gene under the control of the constitutive G6PD promoter. This allows investigation of whether *alcA* or *aldA* is directly subject to carbon-catabolite repression (as well as indirectly via the action of *creA* on *alcR* expression) and the localization of DNA sequences necessary for carbon-catabolite repression. The results (M. Devchand, personal communication) suggest that *alcA* is directly subject to carbon-catabolite repression, but that the sequences responsible for this are upstream of and separable from the part of the promoter necessary for *alcR*-dependent transcriptional control.

2. Secretion and Translation

Very little is known about the molecular biology of secretion in filamentous fungi. In yeast expression-secretion systems, either secretion or translation is usually the rate-limiting process in extracellular protein production.

In filamentous fungi, we expected the secretion capacity to be significantly greater, but secretion could still become limiting at high levels of protein synthesis. In order to investigate the components of either process effectively, it was important to have a homologous in-vitro translation system that can be coupled to homologous or heterologous endoplasmic reticulum membrane preparations to produce an in-vitro translation-translocation system for *Aspergillus*. Building on the work of Devchand and Kapoor (1984) on *N. crassa* cell-free translation, we modified (Devchand et al., 1988) the preparation and optimized conditions of micrococcal nuclease digestion and levels of Mg^{2+} and K^+. After homogenization to disrupt cell walls, GTP is added to 0.1 mM and amino acid pools are depleted by passage of the 23,000 g supernatant through Sephadex G-25. After the optimal micrococcal nuclease treatment, adjustment of the K^+ and Mg^{2+} levels, and the addition of amino acids and an energy-regenerating system, exogenous RNA is efficiently translated. The *N. crassa* lysate translated *A. niger* mRNA encoding high MW proteins more efficiently than a rabbit reticulocyte lysate (NEN). Subsequently, we were able (Devchand et al., 1989) to repeat the preparation of a fungal cell-free translation system using *A. nidulans*. When added to this *A. nidulans* translation system, microsomal membranes prepared from *A. nidulans* were able to direct the translocation of yeast prepro-α-factor into the endoplasmic reticulum. Thus we have laid the groundwork for a concerted genetic and biochemical attack on the molecular basis of translation and secretion in *A. nidulans*.

The presently most accessible component of the secretion mechanism is the signal peptide. Another group in the Allelix laboratory asked whether the quantitative efficiency of various signal peptides could be correlated with amino-acid sequence or protein structure. Experiments in which a variety of yeast signal peptides were placed in the same promoter and coding region context by replacement in-vitro mutagenesis showed that there was a repeatable pattern of variation among signal peptides directing in-vivo secretion, but that the differences were only relevant at low levels of expression and copy number (M. Sutherland, R. Bozzato, and N. Skipper, personal communication). A series of artificial signal sequences were synthesized to test various hypotheses about the roles of amino acid sequence and protein secondary structure in signal sequence function. Some of these performed less well than natural sequences in directing protein secretion at low levels of expression; some actually performed significantly better. However, all signal peptides that maintained a reasonable level of efficacy performed identically at high levels of expression (M. Sutherland and R. Bozzato, personal communication). We concluded from this work that, providing that the signal peptide used was reasonably capable of directing protein secretion in filamentous fungi, optimization of the signal peptide sequence would not be relevant to the yield of secreted protein as long as transcription and translation were efficient.

VIII. FUTURE DIRECTIONS

Genetic engineering of expression and secretion in *A. nidulans* has hardly begun, and several major areas of fungal molecular biology must be investigated. Fortunately, it seems that principles, and many of the molecular details, established in yeast and even in higher eukaryotic cells are applicable to basic molecular systems of filamentous fungal cells. Where should emphasis be placed? The greatest practical short-term rewards will probably come from an investigation of the sequence elements, mRNA binding proteins (Richter, 1988), mRNA processing, and degrading enzymes involved in setting the levels of mRNA translation efficiency and stability in fungal cells. The protein yields obtained rarely reflect the very high level of mRNA from *alcA* promoter constructs, and a drop in mRNA level of a chimeric *alcA* tubulin construct was observed (Waring et al., 1989). In yeast and higher eukaryotes, it is clear that sequences in the 3′ untranslated region of mRNAs have strong effects on mRNA lifetimes (Brawerman, 1987, 1989), and advantage should be taken of recent advances in knowledge in this area, including the direct borrowing of sequences from other organisms. The 3′ end of constructs has been neglected, except for the insertion of a fungal termination signal, partly because each gene to be expressed brings its own 3′ end sequences with it, and knowledge of the biological role of 3′ untranslated sequences has been too limited until recently to attempt rational modification. Many important genes, such as those for growth factors, give rise to unstable messenger RNAs. It is important to determine, either empirically or by studying the signals for mRNA degradation and polyadenylation, what makes an mRNA stable in fungi and to use that information to modify the transcripts resulting from future constructs. The termination of mRNA synthesis is also an unexplored area; in higher eukaryotes a surprising number of genes terminate a significant proportion of their transcription a few hundred nucleotides into the transcription unit (Proudfoot, 1989). There are a series of similar points where the molecular mechanisms controlling mRNA lifetime and translational efficiency are beginning to be elucidated, and those committed to the use of fungal systems for gene expression must continually investigate whether these phenomena and mechanisms occur in fungi and take advantage of every opportunity to improve the stability of the target mRNA and to direct its preferential translation.

Improvement in the rate of mRNA synthesis per gene copy of *alcA*-driven constructs should be possible in the near future. If, by analogy with other systems, it binds sequence C in Figure 2 and activates RNA polymerase directly or indirectly, the identification of the amino acids and nucleotides involved in these interactions will allow a semidirected approach to enhancing the rate of transcriptional initiation. It should also be noted that a yeast-viral hybrid control protein in which a portion of a yeast positive activator protein (gal4) provided recognition specificity, while the viral VP16 protein provided high affinity for RNA polymerase II, was capable of directing higher levels of transcription than the native yeast control protein (Ptashne, 1988). There is no reason why such an approach should not work in the *alcA/alcR* system of *A. nidulans* if *alcR* protein functions in an analogous manner to gal4 of yeast. The chromosomal context of the expression cassettes in transformants may also be worth addressing, particularly if a simple and reliable homologous replacement system can be set up (using *sC* or other genes) that allows the targeting of tandem integrated arrays of expression cassettes preferentially to a particular region.

Present knowledge allows the development of strains of *A. nidulans* in which the *alcA-alcR* system is not susceptible to glucose repression, which would have the advantage of allowing the synthesis of product to take place during growth on the optimal carbon source and would enable the development of continuous fermentation processes in addition to the present batch approach. If present evidence holds up, *alcA* promoters with 320 nucleotides upstream of the translation start are only susceptible to glucose repression indirectly via control of *alcR* expression; therefore, if a host strain were constructed with *alcR* under the control of a different promoter that is not susceptible to carbon catabolite repression (constitutive promoters or inducible promoters from genes not involved in carbon source metabolism), the products of *alcA* promoter-driven genes would be synthesized in the presence of glucose (and inducer). Such strains would retain all the desirable properties of the present system, since it has been shown that increased levels of *alcR* product do not result in an increase in the basal (uninduced) level of transcription of *alcA* (Lockington et al., 1987).

The remaining areas with potential for improvement require much more basic research before clear-cut applications to expression-system improvement can be devised. Secretion will at some point become limiting, while many proteins cannot be secreted efficiently. In this field, there are two major ways to go: modification of the sequence of proteins to be secreted, in order to confer greater flexibility (and then secretability) or enhanced recognition by proteins that modulate the rate of folding of newly synthesized proteins; and modification of the availability of key proteins of the secretory apparatus, such as folding-modulating proteins and the SRP receptor (Meyer, 1988). It should now be possible to begin to identify such proteins in *Aspergillus* or to incorporate genes from other systems encoding such functions. The final area, that of the qualitative and quantitative effects on cellular physiology of the high-level synthesis of foreign mRNA and protein, remains a black box, which will probably only be investigated in the context of the long-term commercial production of particular commercially valuable proteins in a fungal system.

IX. CONCLUSIONS

1. Expression vectors based on the *alcA* promoter of *Aspergillus nidulans*, in combination with host strains carrying multiple copies of the *alcR*-positive activator gene, allow the construction of strains that are useful industrial sources of a variety of commercially valuable proteins.
2. Recombinant constructs are stably maintained as multiple copies in an integrated form in the host genome. The system has a very low basal level, even in the presence of high levels of *alcR* product, and is rapidly inducible by simple substances.
3. The inclusion of an amino-terminal signal peptide sequence allows efficient high-level secretion of any protein preadapted to passage through the secretory pathway.
4. Transformation, selection, and expression systems have been developed that allow the use of *A. niger* glucoamylase promoter-based constructs, *A. niger* acid phosphatase promoter-based constructs, or potentially the *A. nidulans alcA-alcR* system in *Aspergillus niger*, should this species be preferred as a host organism. Proteins can be produced either intra- or extracellularly.
5. The *trans*- and *cis*-acting elements involved in the control of *alcA* expression have been defined, allowing molecular level correlates of industrial performance to be monitored. Together with the development of in-vitro protein translation and translocation systems for *A. nidulans*, this provides an important base for further improvement of the performance of the system.
6. The use of molecular biological knowledge and technology has produced industrial strains of *A. nidulans* within a timeframe easily competitive with traditional approaches, with the advantage that each strain is precisely defined genetically and biochemically. The success of this work should signal the end of the mythological era of industrial mycology.

ACKNOWLEDGMENTS

The largest portion of this work was carried out by a team which I assembled at Allelix Inc., Mississauga, Canada, building on work in my and Dr. C. Scazzocchio's laboratories at the University of Essex, Department of Biology, (with particular contributions from Drs. R.A. Lockington and H.M. Sealy-Lewis) and the University of Manchester, Institute of Science and Technology, Department of Biochemistry and Applied Molecular Biology, both in the United Kingdom. The Allelix laboratory collaborated on *alcA* and *aldA* with Drs. B. Felenbok and D. Sequeval in Professor Scazzocchio's section of the Institut de Microbiologie, Université de Paris XI, Orsay, France, and on acid phosphatase genes with Professor H.N. Arst and Dr. M.X. Caddick of the Department of Microbiology, Royal Postgraduate Medical School, University of London, United Kingdom. I thank all these institutions and individuals for their support, and Professors Scazzocchio and Arst for discussions. Particular recognition is due to Drs. F.P. Buxton and D.I. Gwynne for their major contributions to the success of the work at Allelix Inc. and to others who played important parts: Drs. R. Barnett, M. Devchand, Z.-M. Guo, W.D. MacRae, M. Pickett, and M. Saunders, with S.A. Williams, S. Garven, S. Sibley, R. Elliott, D. Drake, and H. Erfle. I thank Drs. J. Rossant, A. Joyner, A. Bernstein, and L. Siminovitch of Mount Sinai Hospital Research Institute, Toronto, for providing office (and laboratory) facilities during the preparation of this manuscript, and for hospitality. The author is presently Robertson Professor of Biotechnology in the Robertson Institute of Biotechnology, University of Glasgow, Department of Genetics, Scotland, United Kingdom.

REFERENCES

Arst, H.N. Jr. (1981). Aspects of the control of gene expression in fungi. *Symp. Soc. Gen. Microbiol. 31*: 131-160.

Bailey, C. and Arst, H.N. Jr. (1975). Carbon catabolite repression in *Aspergillus nidulans. Eur. J. Biochem. 51*: 573-577.

Ballance, D.J., Buxton, F.P., and Turner, G. (1983). Transformation of *Aspergillus nidulans* by the orotidine-5'-phosphate decarboxylase gene of *Neurospora crassa. Biochem. Biophys. Res. Commun. 112*:284-289.

Barbesgaard, P. (1977). Industrial enzymes produced by members of the genes *Aspergillus*. In *Genetics and Physiology of Aspergillus*, Smith, J.E. and Pateman, J.A., eds. Academic Press, New York.

Berse, B., Dmochowska, A., Skrzypek, M., Weglenski, P., Bates, M.M., and Weiss, R.L. (1983). Cloning and characterisation of the ornithine carbamoyl transferase gene from *Aspergillus nidulans. Gene 25*: 109-117.

Brawerman, D. (1987). Determinants of messenger RNA stability. *Cell 48*:5-6.

Brawerman, D. (1989). mRNA decay: finding the right targets. *Cell 57*: 9-10.

Buxton, F.P., Gwynne, D.I., and Davies, R.W. (1985). Transformation of *Aspergillus niger* using the *argB* gene of *Aspergillus nidulans*. *Gene 37*: 207-214.

Buxton, F.P., Gwynne, D.I., Garven, S., Sibley, S., and Davies, R.W. (1987). Cloning and molecular analysis of the ornithine carbamoyl transferase gene of *Aspergillus niger*. *Gene 60*: 255-266.

Buxton, F.P., Gwynne, D.I., and Davies, R.W. (1989). Cloning of a new bidirectionally selectable marker for *Aspergillus* strains. *Gene 84*: 329-334.

Caddick, M.X. and Arst, H.N. Jr. (1987). Structural genes for phosphatases in *Aspergillus nidulans*. *Genet. Res. Camb. 47*: 83-91.

✓ Caddick, M.X., Brownlee, A.G., and Arst, H.N. Jr. (1987a). Regulation of gene expression by pH of the growth medium in *Aspergillus nidulans*. *Molec. Gen. Genet. 203*: 346-353.

Caddick, M.X., Brownlee, A.G., and Arst, H.N. Jr. (1987b). Phosphatase regulation in *Aspergillus nidulans*: responses to nutritional starvation.

Capecchi, M.R. (1989). The new mouse genetics: altering the genome by gene targeting. *Trends Genet. 5*: 70-76.

Creaser, E.H., Porter, R.L., Britt, K.A., Pateman, J.A., and Day, C.K. (1984). Purification and preliminary characterization of alcohol dehydrogenase from *Aspergillus nidulans*. *Biochem. J. 225*: 449-454.

Devchand, M., Buxton, F.P., Gwynne, D.I., and Davies, R.W. (1988). Preparation of a cell-free translation system from a wild-type strain of *Neurospora crassa*. *Curr. Genet. 13*: 323-326.

Devchand, M., Gwynne, D.I., Buxton, F.P., and Davies, R.W. (1989). An efficient cell-free translation system from *Aspergillus nidulans* and *in vitro* translocation of pre-pro-α-factor across *Aspergillus* microsomes. *Curr. Genet. 14*: 561-566.

Devchand, M. and Kapoor, M. (1984). Preparation of a cell-free translation system from a wild-type strain of *Neurospora crassa* and translation of pyruvate kinase messenger RNA. *Biosci. Rep. 4*: 979-986.

✓ Felenbok, B., Sequeval, D., Mathieu, M., Sibley, S., Gwynne, D.I., and Davies, R.W. (1988). The ethanol regulon in *Aspergillus nidulans*: characterization and sequence of the positive regulatory gene *alcR*. *Gene 73*: 385-396.

Fish, E.N., Banerjee, K., and Stebbing, N. (1983). Human leukocyte interferon subtypes have different antiproliferative and antiviral activities on human cells. *Biochem. Biophys. Res. Commun. 112*: 537-546.

Gwynne, D.I., Buxton, F.P., Sibley, S., Davies, R.W., Lockington, R.A., Scazzocchio, C., and Sealy-Lewis, H.M. (1987a). Comparison of *cis*-acting control regions of two co-ordinately controlled genes involved in ethanol utilisation in *Aspergillus nidulans*.

Gwynne, D.I., Buxton, F.P., Williams, S.A., Garven, S., and Davies, R.W. (1987b). Genetically engineered secretion of bacterial and mammalian proteins from *Aspergillus nidulans*. *Bio/Technology 5*: 713-719.

Gwynne, D.I., Buxton, F.P., Gleeson, M.A., and Davies, R.W. (1987c). Genetically engineered secretion of foreign proteins from *Aspergillus* species. In *Protein Purification: Micro to Macro*, Alan R. Liss, New York, pp. 355-365.

Gwynne, D.I., Buxton, F.P., Williams, S.A., Sills, M., Johnstone, J.A., Buch, J.K., Guo, Z-M., Drake, D., Westphal, M., and Davies, R.W. (1989). Development of an expression system in *Aspergillus nidulans*. *Biochem. Soc. Trans. 17*: 338-340.

Kelly, J.M. and Hynes, M. (1985). Transformation of *Aspergillus niger* by the *amdS* gene of *Aspergillus nidulans*. *EMBO J. 4*: 475-479.

Kozak, M. (1984). Compilation and analysis of sequences upstream from the translational start site in eukaryotic mRNAs. *Nucleic Acids Res. 12*: 857-872.

Kozak, M. (1986). Point mutations define a sequence flanking the AUG initiator codon that modulates translation by eukaryotic ribosomes. *Cell 44*: 283-292.

Kunkel, T.A. (1985). Rapid and efficient site-specific mutagenesis without phenotypic selection. *Proc. Natl. Acad. Sci. USA 82*: 488-492.

Lockington, R.A., Sealy-Lewis, H.M., Scazzocchio, C., and Davies, R.W. (1985). Cloning and characterisation of the ethanol-utilisation regulon of *Asperigillus nidulans*. *Gene 33*: 137-149.

Lockington, R.A., Scazzocchio, C., Sequeval, D., Mathieu, M., and Felenbok, B. (1987). Regulation of *alcR*, the positive regulatory gene of the ethanol utilisation regulon of *Aspergillus nidulans*. *Molec. Microbiol. 1*: 275-281.

MacRae, W.D., Buxton, F.P., Sibley, S., Garven, S., Gwynne, D.I., Davies, R.W., and Arst, H.N. Jr. (1988). A phosphate-repressible phosphatase gene from *Aspergillus niger*: its cloning, sequencing and transcriptional analysis. *Gene 71*: 339-348.

MacRae, W.D., Buxton, F.P., Gwynne, D.I., and Davies, R.W. (1989). Regulated expression and secretion of human interferon α2 in *Aspergillus niger* and *nidulans* using the *Aspergillus niger* acid phosphatase promoter and signal peptide sequence. *Gene,* in press.

Meyer, D.I. (1988). Protein conformation: the year's major theme in translocation studies. *Trends Biochem. Sci. 13*: 471-474.

Pickett, M., Gwynne, D.I., Buxton, F.P., Elliott, R., Davies, R.W., Lockington, R.A., Scazzocchio, C., and Sealy-Lewis, H.M. (1987). Cloning and characterisation of the *aldA* gene of *Aspergillus nidulans*. *Gene 51*: 217-226.

Proudfoot, N. (1989). How RNA polymerase II terminates transcription in higher eukaryotes. *Trends Biochem. Sci. 14*: 105-110.

Ptashne, M. (1988). How eukaryotic transcriptional activators work. *Nature 335*: 683-689.

Richter, J.D. (1988). Information relay from gene to protein—the mRNP connection. *Trends Biochem. Sci. 13*: 483-486.

Skipper, N., Bozzato, R.P., Vetter, D., Davies, R.W., Wong, R., and Hopper, J.E. (1987). Use of the melibiase promoter and signal peptide to express a bacterial cellulase from yeast. In: *Biological Research and Industrial Yeasts*, Stewart, G.G., Russell, I., Klein, R.D., and Hiebsch, R.R., eds., CRC Press, Boca Raton, FL, p. 137-147.

Skipper, N., Sutherland, M., Davies, R.W., Kilburn, D.G., Miller, R.C., Warren, R.A.J., and Wong, R. (1985). Secretion of a bacterial cellulase by yeast. *Science 230*: 958-960.

Slocombe, P., Easton, E., Boseley, P., and Burke, D.C. (1982). High expression of an interferon α2 gene cloned in phage M13mp7 and subsequent purification with a monoclonal antibody. *Proc. Natl. Acad. Sci. USA 79*: 5455-5459.

Sealy-Lewis, H.M. and Lockington, R.A. (1984). Regulation of two alcohol dehydrogenoses in *Aspergillus nidulans*. *Curr. Genet. 8*: 253-259.

Tilburn, J., Scazzocchio, C., Taylor, G.G., Zabicky-Zissman, J.M., Lockington, R.A., and Davies, R.W. (1983). Transformation by integration in *Aspergillus nidulans*. *Gene 26*: 205-221.

Waring, R.B., May, G.S., and Morris, N.R. (1989). Characterization of an inducible system in *Aspergillus nidulans* using *alcA* and tubulin-coding genes. *Gene 79*: 119-130.

Yelton, M.M., Hamer, J.E., and Timberlake, W.E. (1984). Transformation of *Aspergillus nidulans* using a *trpC* plasmid. *Proc. Natl. Acad. Sci. USA 81*: 1370-1374.

Von Heijne, G. (1986). A new method for predicting signal sequence cleavage sites. *Nucleic Acids Res. 14*: 4683-4690.

4

Chymosin Production in *Aspergillus*

Michael Ward
Genencor International, South San Francisco, California

I. INTRODUCTION

Other chapters in this volume have provided an overview of molecular genetics and heterologous gene expression in filamentous fungi. This chapter will deal with the production of bovine chymosin by *Aspergillus* as a specific example of the application of these techniques.

Chymosin (rennin; EC 3.4.23.4) is an aspartyl protease extracted from the abomasum (fourth stomach) of unweaned calves and used in cheese manufacture. It is initially secreted as a zymogen, prochymosin (365 amino acid residues), from which a 42-residue, amino-terminal prosequence, rich in basic amino acids, is autocatalytically removed at low pH to yield active, mature chymosin (Pedersen et al., 1979). Milk clotting is achieved by a fairly specific cleavage of $\varkappa$-casein by chymosin (Foltman, 1979). The nucleotide sequence of chymosin cDNA clones has shown that the initial translation product is preprochymosin, having a 16 amino acid secretion-signal sequence at the amino terminus (Moir et al., 1982; Harris et al., 1982; Nishimori et al., 1982a). Two forms of chymosin, A and B, differing by a single amino acid substitution, are encoded by alleles of the same bovine gene (Foltmann et al., 1977; Foltmann et al., 1979).

Chymosin has significant amino acid sequence homology with other aspartyl proteases (see Chapter 5) such as pepsin (Tang et al., 1973; Tang, 1979) and fungal aspartyl proteases, including aspergillopepsin of *Aspergillus awamori* (Stepanov, 1985), endothiapepsin from *Endothia parasitica* (Barkholt, 1987) and *Mucor* rennins (Tonuochi et al., 1986; Gray et al., 1986). Indeed, due to a shortage of authentic calf chymosin, both the *Endothia* and *Mucor* enzymes are used as substitutes for chymosin in cheese manufacture. However, these proteases lack the specificity of chymosin and are not ideal.

There have been several reports of the production of calf chymosin in microorganisms as a result of recombinant DNA technology. Initial attempts were made with *Escherichia coli* as the host organism (Nishimori et al., 1982b; Emtage et al., 1983), but, although high levels of intracellular expression were achieved, the chymosin was deposited as insoluble inclusion bodies (Beppu, 1983; Emtage et al., 1983; Kawaguchi et al., 1987). In order to recover active chymosin in significant quantities from these cells, it was necessary to purify the inclusion bodies and to treat them with urea at high pH to solubilize and refold the chymosin (Emtage et al., 1983; Nishimori et al., 1984; Kawaguchi et al., 1984; Marston et al., 1984; McCaman et al., 1985). The highest reported yield of active enzyme was 30% of the total chymosin (Nishimori et al., 1984), but this strategy is economically unattractive.

The yeast *Saccharomyces cerevisiae* has also been used as host for chymosin production (Mellor et al., 1983; Goff et al., 1984; Moir et al., 1985). Again the intracellular chymosin produced was found to be in an insoluble and inactive form in this system (Mellor et al., 1983; Moir et al., 1985). The natural chymosin secretion signal was not able to direct secretion by *S. cerevisiae*, but the addition of the *S. cerevisiae* invertase secretion signal to prochymosin allowed the secretion of 10% of the total prochymosin made (Moir et al., 1985). The secreted prochymosin could be activated to chymosin by low pH in the same way as natural prochymosin. Subsequent use of "supersecretor" mutant strains of *S. cerevisiae* resulted in secretion of up to 50% of the prochymosin produced, although the total amount remained constant (Smith et al., 1985). An additional advantage to secretion of chymosin is the greater ease with which it can be purified away from cellular material.

More recently, other yeast species have been used as hosts for chymosin production. With *Kluyveromyces lactis* (van den Berg et al., 1990), intracellular chymosin was found to be soluble and active, and, if a secretion signal was used (e.g., the native preprochymosin or a yeast signal sequence), more than 90% of the prochymosin was found in the medium. The efficiency of secretion from this host was thus in marked contrast to that of *S. cerevisiae.*

In addition to yeasts, filamentous fungi have also been used as hosts for chymosin production. This chapter will deal mainly with production in *Aspergillus nidulans* and *A. awamori*, although recently *Trichoderma reesei* has also been utilized as a host organism (see Chapter 6; Harkki et al., 1989). Some of the advantages inherent to either the prokaryote or yeast hosts mentioned above are shared by the filamentous fungi, including their previous use for the production of native enzymes generally regarded as safe for food use and the fact that the technology required for large-scale fermentation is already in place. A major advantage of filamentous fungi over previously explored hosts is the variety and high levels of naturally occurring secreted proteins that they are capable of producing. In the case of glucoamylase production by *A. niger* and *A. awamori*, this can exceed concentrations of 20 g/l in large-scale fermentation vessels. It would obviously be extremely beneficial if this potential could be realized for the secretion of heterologous proteins. A further advantage, more specific to chymosin production, is that chymosin is very similar to the aspartyl protease produced as a native secreted enzyme by *A. awamori* and several other species. This latter advantage of similarity between the mammalian enzyme to be produced and a native enzyme is, however, offset somewhat by the ability of the native aspartyl protease to degrade the desired recombinant product.

There have been several reports of the production of heterologous proteins in the genetically well-characterized species, *A. nidulans*, with a variety of fungal promoters used to direct the expression of the foreign gene (Cullen et al., 1987a; Gwynne et al., 1987; Upshall et al., 1987; Turnbull et al., 1989). These studies are now being extended to those industrially important species capable of secretion of copious amounts of native protein, such as *A. niger*, *A. oryzae* (Christensen et al., 1988), and *T. reesei* (Harkki et al., 1989).

II. CHYMOSIN PRODUCTION IN *ASPERGILLUS NIDULANS*

A. Expression from the *glaA* Promoter

Initial experiments at Genencor on the production of chymosin from filamentous fungi were conducted with *A. nidulans*. Although this species is not noted for the secretion of prodigious amounts of native proteins, it is genetically well characterized and transformation systems were already well established at the onset of this work. In order to direct the expression of a chymosin B cDNA clone, the promoter from a highly expressed *Aspergillus* gene and a fungal secretion signal sequence were sought. To supply these sequences the *A. niger* glucoamylase gene (*glaA*) (Boel et al., 1984) was chosen. Several plasmid constructions were made incorporating the *glaA* gene promoter and terminator, prochymosin cDNA, and optional use of the secretion signals from either gene or part of the glucoamylase coding sequence (Figure 1). Introduction of these plasmids into *A. nidulans* (strain G191: *pabaA1*, *pyrG89*; *fwA1*, *uaY9*) by transformation, using the *Neurospora crassa pyr4* gene as a selectable marker, allowed the production and secretion of active chymosin (Cullen et al., 1987a). Western blots probed with anti-chymosin antibody confirmed that the chymosin produced was of similar size to authentic calf chymosin. All of the constructions shown in Figure 1 were capable of directing chymosin production, pGRG1 giving the highest levels (up to 159 μg/g mycelial dry weight, approximately 3 μg/ml, in a 50-ml shake-flask culture of transformant GRG1-1) and pGRG2 giving the lowest levels of production. Unlike *S. cerevisiae*, *A. nidulans* was able to utilize the mammalian secretion signal (present in pGRG3) with similar efficiency as the fungal secretion signal, suggesting a less stringent requirement for this sequence in the filamentous fungus. Significantly, not only was the *A. niger* promoter functional in *A. nidulans*, as might be expected from other work involving the transfer of genes between filamentous Ascomycetes (Ballance, 1986), but expression was under similar regulation in the heterologous host. Thus levels of chymosin-specific mRNA and chymosin production were minimal if xylose was used as a carbon source for the growth of the transformed *A. nidulans* strains, whereas expression was maximal in the presence of starch. Glucoamylase production in *A. niger* is regulated in a similar manner. Although the level of chymosin production was low in these transformations, there was apparently little intracellular accumulation of chymosin, suggesting that the majority (more than 90%) of the chymosin produced was secreted. Southern blotting and hybridization of DNA isolated from transformants suggested that the level of chymosin production was not correlated with the copy number of intact chymosin expression cassettes integrated into the genome.

B. Expression from the *oliC* Promoter

In addition to the *glaA* gene promoter used above, a promoter from a highly expressed, constitutive *A. nidulans* gene has also been tested. The

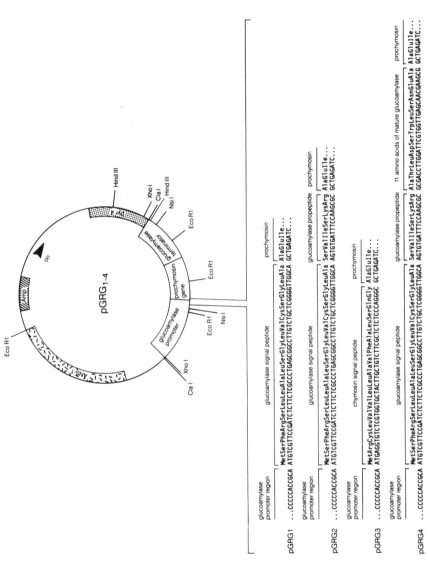

Figure 1 Restriction map of plasmids pGRG1-4 showing nucleotide sequences of the glucoamylase-chymosin fusions. (From Cullen et al., 1987a, with kind permission of the publishers.)

gene in question was the *oliC* gene (Ward et al., 1986; Ward and Turner,
1986), which encodes subunit 9 of the mitochondrial ATP synthase complex. In *N. crassa*, subunit 9 constitutes approximately 0.2-0.4% of the
total cellular protein, and cDNA cloning experiments suggest that the message level is about 0.2% of the total poly A^+ RNA (Viebrock et al., 1982).
This gene should be expressed at the highest level during periods of maximal growth, whereas the *A. niger glaA* gene is not fully expressed until
later in the log phase of growth of a shake-flask culture. These expectations have been confirmed by Northern blot hybridization analysis of RNA
extracted at various time points during shake-flask culture. The construction used for this work (pOR1) was essentially the same as pGRG1, but
with the *oliC* promoter exchanged for the *glaA* promoter (Figure 2).
In order to construct pOR1, a *bclI* site was created by in vitro mutagenesis at the junction between the promoter and coding sequences of both
the glucoamylase and *oliC* genes (Figure 2). This mutagenesis left the *oliC*
promoter unaltered, but changed the second and third amino acids of the
glucoamylase secretion signal from Ser Phe to Ile Ile. The *ans1* sequence,
which increases the transformation frequency in *A. nidulans* (Ballance
and Turner, 1985), was not included in pOR1. Following transformation
into the *A. nidulans* strain G191, the plasmid pOR1 was capable of directing the expression and secretion of active chymosin. To measure chymosin
activity, a microtiter plate assay was developed based on the increase in
turbidity resulting from milk clotting. Samples were diluted in 10 mM sodium phosphate buffer at pH 6.0 and added to substrate (1% skin milk,
40 mM $CaCl_2$, 50 mM sodium acetate, pH 6.0). After incubation for 15
min at 37 °C, turbidity was measured at 690 nm. In addition, an enzyme
immunoassay (EIA) utilizing rabbit anti-chymosin antibody was used to
measure the total amount of chymosin protein (Cullen et al., 1987a), which
was generally higher than the results obtained from activity assays, since
it includes inactive prochymosin and degraded chymosin.

Several transformants obtained with pOR1 were screened for activity
and the highest chymosin producer (OR1-10) was chosen for further study.
To optimize expression levels in 50-ml shake-flask cultures, the transformants were grown in SCM consisting of *Aspergillus* minimal medium (Rowlands and Turner, 1973) with 2% malt extract, 0.5% yeast extract, 0.1%
peptone, 0.1% Tween 80, and 1 ml/l Mazu DF60-P antifoam. In addition,
100 mM $(NH_4)_2SO_4$ was added to repress the production of native *Aspergillus* extracellular proteases (Cohen, 1977), which otherwise cause degradation
of chymosin.

Northern blotting analysis was employed to study the relative abundance
of *oliC* mRNA and chymosin mRNA in transformant OR1-10 (Figure 3).

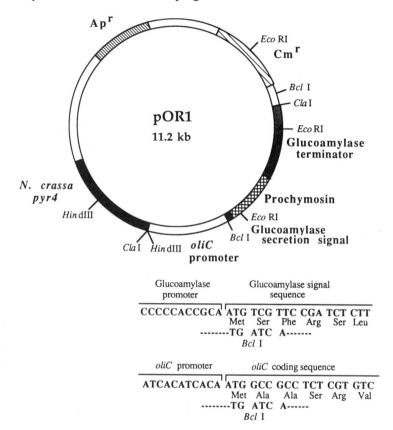

Figure 2　Restriction map of plasmid pOR1. The sequences around the translation initiation sites of the glucoamylase and *oliC* genes are shown, including the changes made by in vitro mutagenesis to create *BclI* sites.

The DNA fragments used as probes to detect the two different mRNA species were of a similar length and were nick translated to similar specific activities so that these comparisons could be made. The steady-state level of chymosin mRNA produced from the *oliC* promoter was apparently much lower than the level of *oliC* message produced from the native gene. Comparatively low levels of heterologous mRNA were also obtained in *A. nidulans* with a translational fusion in which the *oliC* promoter and the first six codons on the *oliC* coding sequence were fused to the *E. coli* *lacZ* gene (S. Kerry-Williams, M. Ward, and G. Turner, unpublished results). A similar phenomenon has been observed in many attempts at the

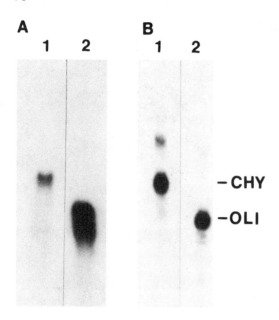

Figure 3   Northern blots of 10 μg of total RNA from *A. nidulans* transformants hybridized with nick-translated DNA fragments specific for chymosin or *oliC* mRNA. A: transformant OR1-10. B: transformant GRG1-1. Lane 1 hybridized with the chymosin-specific probe and lane 2 hybridized with the *oliC*-specific probe in each case.

expression of heterologous genes in *S. cerevisiae* (see Chen et al., 1984 for discussion). Chymosin mRNA levels were also low in *T. reesei* using the *cbh1* gene promoter to direct expression (see Chapter 6; Harkki et al., 1989). This phenomenon may be due to a shorter half-life of the heterologous message or reduced rates of transcription from the recombinant DNA construct compared to the native gene. In contrast, levels of mRNA derived from a fusion between the *E. coli* enterotoxin subunit B gene and the *amdS* promoter were greater than the levels of native *amdS* message in *A. nidulans* (Turnbull et al., 1989). Similarly, abundant chymosin-specific mRNA is obtained using the *glaA* promoter in *A. awamori* (see below).

It is also apparent from Figure 3 that the level of chymosin-specific message in transformant GRG1-1 (*glaA* promoter construct) was similar to the level of native *oliC* message and was therefore significantly higher than the steady-state level of chymosin mRNA in transformant OR1-10 (*oliC* promoter construct).

The results of an experiment designed to study the time course of chy- mosin production in shake-flask cultures of transformants GRG1-1 and OR1-10 are shown in Figure 4. The dry weight curves indicate that the two cultures grew at comparable rates. Although chymosin reached a similar peak concentration of approximately 2 μg/ml in the two cultures, production began earlier in the culture of OR1-10, as would be expected from the temporal difference in the expression of the *glaA* and *oliC* genes. It is interesting to note that the yield of chymosin from the two transformants was similar, despite the fact that transformant GRG1-1 could produce much higher steady-state levels of chymosin mRNA. This, together with the fact that the plasmid copy number does not correlate with production levels, suggests that message levels are not the limiting factor for chymosin production but that some other process, possibly involved with translation

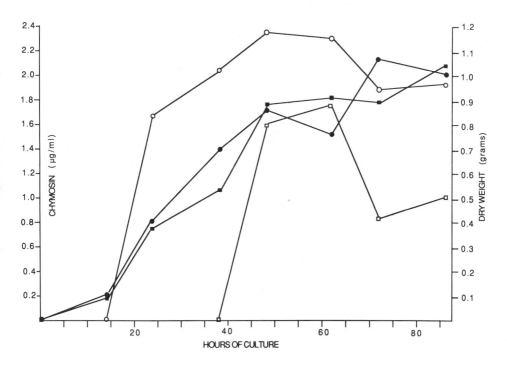

Figure 4 Chymosin production and growth rates of transformants OR1-10 and GRG1-1. A single 50 ml shake-flask culture of each transformant was harvested at the times indicated, and chymosin activity and dry weight were measured. □———□, GRG1-1 chymosin concentration; ■———■, GRG1-1 dry weight; ○———○, OR1-10 chymosin concentration; ●———●, OR1-10 dry weight.

or secretion, is limiting. It might be interesting to insert both the *oliC* and the *glaA* promoter constructions into a single transformed strain so that chymosin is produced both during early and late growth periods.

III. CHYMOSIN PRODUCTION IN *ASPERGILLUS AWAMORI*

A. Molecular Genetics of *A. awamori*

In contrast to *A. nidulans*, strains of the closely related species *A. niger* and *A. niger* var. *awamori* (*A. awamori*) are poorly characterized genetically but are capable of secreting large quantities of native enzymes (e.g., glucoamylase, the gene for which is highly conserved between these two species) (Boel et al., 1984; Nunberg et al., 1984). However, transformation systems have recently been reported for *A. niger*, with selection based on complementation of auxotrophic mutations (Buxton et al., 1985; van Hartingsveldte et al., 1987) or the use of dominant selectable markers (Kelly and Hynes, 1985; Ward et al., 1988). It is therefore possible to examine chymosin production from the *glaA* promoter in these species. For this work we have chosen a glucoamylase over-producing strain of *A. awamori* (UVK143f). derived from strain NRRL3112, which is capable of producing approximately 5 g/l glucoamylase in 50 ml shake-flask culture. We have isolated *pyrG* and *argB* auxotrophic strains by methods similar to those used for *A. niger* (Buxton et al., 1985; van Hartingsveldte et al., 1987), and these can be complemented by the *N. crassa pyr4* and *A. nidulans pyrG* genes or the *A. nidulans argB* genes, respectively. Transformation levels are generally low in *A. awamori* and *A. niger* compared to *A. nidulans*, but frequencies of 1-10 transformants per microgram of plasmid DNA can be reproducibly obtained. In addition, we have constructed vectors that allow the selection of transformants on the basis of resistance to hygromycin, G418, or bleomycin. These antibiotics have been used previously for the selection of filamentous fungi transformed with the respective bacterial resistance genes; the hygromycin phosphotransferase gene (Cullen et al., 1987b; Punt et al., 1987), the aminoglycoside phosphotransferase-3' gene (Revuelta and Jayaram, 1986), or the bleomycin-resistance gene (Kolar et al., 1987). In order to allow expression in *A. awamori*, we have constructed vectors in which the *A. nidulans oliC* promoter was used to direct the expression of the bacterial genes. Transformation with these dominant selectable markers is generally less efficient than by the complementation of auxotrophic mutations. Results from a large number of transformation experiments have allowed the following conclusions to be drawn. Circular plasmids, such as the pGRG series, which only have *glaA* sequences in common with the *A. awamori* genome, integrate at the homologous site in the genome extremely rarely. Multiple copies, tandem integration, and multiple integration sites of plasmid are common. Linear fragments of DNA with ends homologous to a region in the genome will transform and integrate at the region of homology in approximately 20-50% of transformants. The latter observation allows gene deletion or replacement experiments to be undertaken, and a strain has been generated in which the native glucoamylase gene has been replaced with the *argB* gene using the strategy outlined in Figure 5. This gene replacement was performed using a double auxotrophic strain (GC12; *pyrG*; *argB*) obtained from a parasexual cross between the two single auxotrophic strains. The resulting glucoamylase deletion strain thus still harbored the *pyrG* mutation and was able to be subsequently transformed with plasmids bearing the *pyr4* or *pyrG* genes. Similar methods have been used to disrupt other genes, including the gene encoding the native aspartyl protease. The above transformation strategies

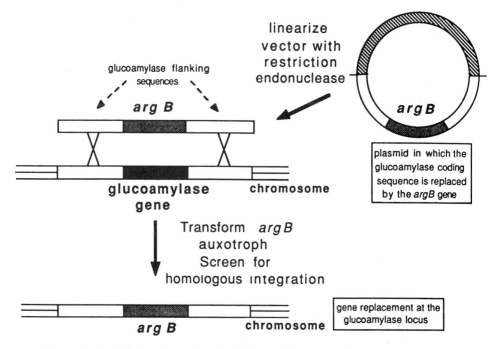

Figure 5   Outline of a scheme for the deletion of the glucoamylase coding sequence from the genome of *A. awamori*.

can now be applied to a wide range of fungal strains and species of industrial and economic importance.

B. Chymosin Production Levels

In order to obtain expression of chymosin in *A. awamori*, transformation of a *pyrG* auxotrophic strain was performed with pGRG1 or pGRG3. A large number of transformants were screened for chymosin production in shake-flask culture and the highest production strain was identified (107, a pGRG3 transformant capable of producing 7-10 μg/ml of active chymosin in shake-flask culture with SCM). Despite the fact that these transformants and the parent strain were capable of producing a large amount of secreted glucoamylase, the level of chymosin production from the *glaA* promoter was not much greater than that obtained from *A. nidulans* (Table 1). Table 1 also shows that glucoamylase production was not reduced in the majority of transformants compared to the parent strain. This observation suggests that there is not a strictly limiting supply of factors involved with transcription, or other aspects of glucoamylase production, for which the chymosin system must compete. A similar conclusion has been drawn from experiments showing that increasing the copy number of the *glaA* gene in *A. niger* by transformation leads to increased levels of glucoamylase production (Finkelstein, 1987). As would be predicted from the above conclusions, the production of chymosin was not noticeably increased in a strain from which the native glucoamylase gene had been

Table 1 Glucoamylase and Chymosin Concentrations in Media from Shake-Flask Cultures of pGRG1 and pGRG3 Transformants of *A. awamori*

Strain	Plasmid	Chymosin (μg/ml)[a]	Glucoamylase (μg/ml)[a]
GC5	None	N.D.	2311
13	pGRG3	4.8	1482
18	pGRG3	5.5	3863
20	pGRG3	4.1	2821
107	pGRG3	14.6	3280
62	pGRG1	4.7	420
82	pGRG1	5.6	3922
83	pGRG1	4.9	3105

N.D. = not detectable.
[a]Average from two cultures after 3 days of growth measured by EIA.

deleted. However, transformant 62 did have a significantly reduced level of glucoamylase production. Southern blotting experiments showed that this transformant had a high number of integrated copies of pGRG1 (results not shown), and it is possible that this DNA titrated out factors required for *glaA* gene expression.

The DNA sequences of the promoter regions of the *A. niger glaA* promoter used in pGRG1 and pGRG3 and the promoter from the *glaA* gene of *A. awamori* are highly homologous to the extent determined (205 bp 5' of the translation initiation codon), suggesting that the *A. niger* promoter should work efficiently in *A. awamori*. However, to check that the *A. awamori* promoter was not significantly different with respect to chymosin production, a plasmid was constructed in which this promoter was exchanged for the *A. niger* promoter in a pGRG4-like construction. The chymosin yields from transformants containing this construction were similar to those obtained from pGRG4 transformants.

Although it was shown that some of the transformants obtained with pGRG1 and pGRG3 had a high number of integrated plasmid copies, strain 107 was found by Southern analysis to have a low plasmid copy number. As in the case of *A. nidulans*, the plasmid copy number did not correlate with high levels of expression of chymosin in *A. awamori*. Northern blots were examined to determine the steady-state levels of chymosin and glucoamylase mRNA in transformant 107.

For this work a probe was constructed that consisted of a single linear fragment of DNA with approximately equal amounts of sequence homologous to each of the two mRNA species. Direct comparison of the amounts of glucoamylase and chymosin message could, therefore, be made from a single blot hybridized with only one probe. It was apparent from these experiments (Figure 6) that the level of chymosin-specific message was slightly lower than that for the native glucoamylase, but this difference was not nearly so pronounced as that between the chymosin mRNA levels derived from the *oliC* promoter and the native *oliC* gene in *A. nidulans*, suggesting that transcription of the chymosin cDNA from the *glaA* promoter was reasonably efficient. Although transcription was apparently not the limiting step in chymosin production in transformant 107, some later step(s) in transcript processing, translation, and/or secretion are obviously much less efficient than for native glucoamylase production.

C. Efficiency of Chymosin Secretion

As noted previously, experiments conducted on *A. nidulans* transformants suggested that more than 90% of the total chymosin produced was secreted to the external medium (Cullen et al., 1987a). Similar experiments were

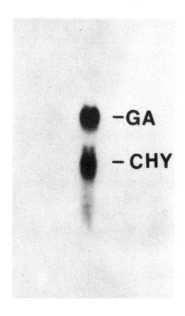

Figure 6   Northern blot of 10 μg of total RNA from transformant 107 hybridized with a single, nick-translated DNA fragment comprising similar lengths of gluco-amylase and chymosin gene sequences.

conducted with the *A. awamori* transformants. All of the mycelium from 50-ml shake-flask cultures was collected and washed by filtration, freeze dried, and ground. This material was resuspended in 50 ml of extraction buffer (50 mM sodium phosphate buffer, pH 5.5 and 500 mM NaCl) containing both 0.1 mM pepstatin and 1 mM phenyl methyl sulphonyl fluoride as protease inhibitors. Extracts were further treated by adding NaOH to a final concentration of 50 mM and incubating for 30 min at 37 °C before centrifugation to remove debris. Chymosin was subsequently assayed by EIA using chymosin-specific antibody. Control experiments using a chymosin standard solution showed that the NaOH treatment reduced the EIA signal by at least 25%. However, this treatment was found to be necessary for solubilization of chymosin from the mycelial extracts. SDS treatment could also be used for this purpose, but was not as effective as NaOH treatment.

   Table 2 gives the results of assays of chymosin concentration in the medium and in mycelium from shake-flask cultures of various pGRG1 or pGRG3 transformants. In this experiment, none of the transformants

Table 2 Intracellular and Extracellular Concentrations of Chymosin in pGRG1 and pGRG3 Transformants of *A. awamori*

Strain	Plasmid	Chymosin concentration (μg/ml)	
		Intracellular	Extracellular
13	pGRG3	14.5	2.5
18	pGRG3	N.D.	2.6
20	pGRG3	14.9	1.7
107	pGRG3	N.D.	7.8
62	pGRG1	33.9	3.0
82	pGRG1	3.4	2.7
83	pGRG1	N.D.	3.6

N.D. = not detectable.

produced more than approximately 8 μg/ml of secreted chymosin as measured by EIA. However, some of them had comparatively high levels of intracellular chymosin (up to approximately 34 μg/ml of extract). Allowing for the decrease in the effective chymosin concentration caused by the NaOH treatment, the likelihood that accumulated intracellular chymosin may be degraded and the possibility that not all the intracellular chymosin could be freed from the mycelial debris, this latter figure is obviously an underestimate. These results indicate that some transformants were able to produce 50 μg/ml or more chymosin (if intra- and extracellular material is included), but secretion of chymosin from these transformants was not efficient. This may also have been true of the *A. nidulans* transformants produced previously (Cullen et al., 1987a), but the extraction methods used at this time (i.e., without pepstatin in the extraction buffer or NaOH treatment) may have been inadequate, or the intracellular chymosin may have been degraded prior to extraction. Transformant 107, selected as the highest producer of secreted chymosin, did not have high intracellular levels. However, the total amount of chymosin (intra- and extracellular) was not as high in this strain as in some of the other transformants. Interestingly, transformant 62, which had the highest internal concentration of chymosin, also had a high plasmid copy number.

D. Mutagenesis of Transformants and Improvement of Host Strain

Repeated rounds of UV light or nitroso-guanidine (NTG) mutagenesis of transformed strain 107, with screening for increased chymosin production after each round, have resulted in considerable yield increases (Lamsa and Bloebaum, 1989). One method of screening for strains that overproduced chymosin involved inoculating mutated spores onto filter discs overlaid on solid complete medium plates containing a colony restrictor (dichloran). After growth of the colonies, the filters were removed and the plates were treated with an aspartyl protease inhibitor, diazoacetyl norleucine methyl ester (DAN; Rajagopalan et al., 1966), which had a differential effect on the native aspartyl protease and chymosin. The concentration of DAN used gave total inhibition of the *Aspergillus* aspartyl protease and partial inhibition of the chymosin. The plates were overlaid with a disc of skim milk agarose (1% agarose, 1% skim milk, 0.45% $CaCl_2$, 0.6% acetic acid, pH 5.5), and milk clotting was allowed to proceed at 37 °C. The extent of clotting was proportional to the amount of chymosin secreted by the colony that had occupied that position on the plate. Comparisons of shake-flask cultures of strain 107 and a strain (2AD73) obtained after four rounds of mutagenesis showed that chymosin production had been increased by approximately fourfold, whereas native glucoamylase production had decreased to approximately 10% of the level of production in strain 107. Northern blot analysis of RNA from mutants obtained from this program showed that there was at least as much chymosin-specific message as glucoamylase message in one of the mutants. The vast majority of the chymosin mRNA was shown, by binding to an oligo (dT) column, to be polyadenylated, demonstrating that this step of message processing had been performed efficiently by this strain.

Improvements have recently been made in the screening procedures for the identification of chymosin overproducing mutants. For example, following mutagenesis spores can be cultured in volumes as low as 200 μl in microtiter dishes, and chymosin activity assays can be performed on the culture supernatant. Many of the manipulations have been automated using laboratory robotics, allowing many more mutagenized spores to be processed. Strains derived from transformant 107 have now been shown to produce up to 100 μg/ml chymosin in shake flasks using improved media. Although no detailed study of the genetics of these mutant strains has been undertaken, it is likely that some of the mutations selected for during this program involved alterations that allowed more efficient translation or secretion of chymosin.

Another mutant strain derived in one step from a different transformant had simultaneously increased chymosin production and decreased native aspartyl protease production to an undetectable level by EIA. A reduction in the latter enzyme may result in decreased degradation of chymosin in the culture supernatant and thus increased production. In support of this suggestion, it was noted that differences between the amount of chymosin measured by EIA and the amount of active chymosin was less marked in cultures of the mutant strain than in the parent strain. The aspartyl protease (aspergillopepsin A) gene has subsequently been cloned and deleted from a strain of *A. awamori* using the strategy outlined in Figure 5 (Ward, 1989a; Berka et al., 1990). As a result the proteolytic activity in culture supernatants was greatly reduced, as judged by zones of clotting on skim-milk agarose plates. Secreted chymosin yields in the deleted strain were approximately twice that obtained from nondeleted strains.

E. Improvement of Chymosin Expression Plasmid

Recently, the development of an expression vector, termed pGAMpR, has led to increased yields of chymosin from *A. awamori* transformants (Ward, 1989b; Ward et al., 1990). This plasmid contains the *A. awamori* glaA promoter and the entire glucoamylase coding sequence fused in frame to the prochymosin B coding sequence. Consequently, a full-length, fused glucoamylase-chymosin transcript and a fused protein were synthesized. During the first few days of culture in medium buffered at a comparatively high pH (pH 6), little mature chymosin was present. However, the fusion protein could be detected by Western blot analysis using antibodies directed against either glucoamylase or chymosin. At later time points during culture, the majority of the chymosin was in the form of mature, active chymosin. It was possible to release chymosin from the fusion protein simply by reducing the pH to 2 for 30 min, suggesting that autocatalytic removal by chymosin of its prosequence and attached glucoamylase may be possible.

Approximately 150 μg/ml of chymosin was produced by a pGAMpR transformant in which the native aspartyl protease gene was intact. Accumulation of intracellular chymosin does not occur in pGAMpR transformants to the extent observed for pGRG transformants, suggesting that secretion of the fusion protein may be more efficient than for chymosin alone. Transformation of a strain deleted for the native aspartyl protease gene with pGAMpR and subsequent mutagenesis and screening has provided still higher yields.

IV. CONCLUSIONS

Many species of filamentous fungi have a high capacity for native protein secretion, and this is a predominant reason for their consideration as hosts for the production and secretion of heterologous proteins. Initial attempts at the expression of nonfungal genes in *A. nidulans* led to low levels (1-10 μg/ml) of secreted product (Cullen et al., 1987; Gwynne et al., 1987; Upshall et al., 1987). Production of chymosin directed by the *cbh1* promoter in *T. reesei* (see Chapter 6; Harkki et al., 1989) yielded somewhat higher levels (approximately 40 μg/ml). Obviously, something was limiting the full potential for production in each case.

The conclusion drawn from the work on chymosin production in *Aspergillus* described in this chapter is that secretion of the mammalian protein was inefficient, although other factors may also have limited productivity. Inefficient secretion was also suggested as a cause for the low extracellular yield of the *E. coli* enterotoxin subunit B obtained in *A. nidulans* (Turnbull et al., 1989) and might be expected to be a problem with other heterologous proteins. The high yields (up to 3.3 mg/ml) of secreted *M. miehei* aspartyl protease that were obtained from *A. oryzae* using the α-amylase gene promoter (Christensen et al., 1988) presumably reflect the fact that the heterologous product was a naturally secreted fungal protein in this case. Attempts to express chymosin in this system resulted in yields similar to those in *A. awamori*, i.e., approximately 10 μg/ml (Boel et al., 1987).

Unfortunately, measurement of intracellular levels of chymosin, or other heterologous protein, is probably not a reliable method for the estimation of secretion efficiency. More sophisticated analysis is required to determine if this is the sole or major limiting step in production. However, without fully understanding the underlying cause of low yields of chymosin, it was possible to increase production by devising a mutagenesis and screening program. This is potentially feasible for any desired product, presuming a method of screening or selecting for overproduction can be developed. A combination of molecular genetics and classical strain improvement techniques can be extremely powerful and will be required until the limitations to production and secretion are understood at the molecular level. Although mutagenesis and screening methods may be laborious, it is comforting to know that in filamentous fungi one is dealing with host systems in which the upper capacity for protein secretion is very high. It may be possible to isolate mutant strains, similar to the "supersecretor" strains of yeast (Smith et al., 1985), which secrete foreign proteins, in general, more efficiently.

Strains of *A. awamori* have now been developed at Genencor that produce commercially viable amounts of secreted chymosin. Extensive biochemical and toxicological tests and cheese-making trials have been conducted with the conclusion that this chymosin constitutes a safe and effective alternative to chymosin obtained from calves.

In addition to chymosin, we have expressed several heterologous fungal genes in *A. nidulans* and *A. awamori*. These include the gene for *M. miehei* aspartyl protease and the *T. reesei* genes for cellobiohydrolase I, cellobiohydrolase II, and endoglucanase I. In some cases expression was obtained using the native *M. miehei* and *T. reesei* promoters, whereas in other experiments these were replaced with the *Aspergillus* glucoamylase promoter. Although yields were again low (approximately 10-50 µg/ml of secreted enzyme), this is an extremely useful system for the expression of altered forms of these genes produced in vitro and for the analysis of the effects of specific changes in protein primary structure on enzyme activity or other function. In each of these studies, it was noted that the heterologous protein was hyperglycosylated when produced in *Aspergillus*. Although the activity of the enzyme may not be noticeably affected by this different glycosylation pattern, the problem obviously deserves further attention.

ACKNOWLEDGMENTS

The previously unpublished work reported in this chapter was performed by a group of scientists at Genencor International, including R.M. Berka, C.C. Barnett, M.W. Rey, S.A. Thompson, L.J. Wilson, K.H. Kodama, K.J. Hayenga, D. Cullen, P. Bloebaum, and M.H. Lamsa.

REFERENCES

Ballance, D.J. (1986). Sequences important for gene expression in filamentous fungi. *Yeast 2*:229-236.

Ballance, D. J., and Turner, G. (1985). Development of a high frequency transforming vector for *Aspergillus nidulans*. *Gene 36*:321-331.

Barkholt, V. (1987). Amino acid sequence of endothiapepsin. Complete primary structure of the aspartic protease from *Endothia parasitica*. *Eur. J. Biochem. 167*:327-338.

Beppu, T. (1983). The cloning and expression of chymosin (rennin) genes in microorganisms. *Trends Biotechnol. 1*:85-89.

Berka, R.M., Ward, M., Wilson, L.J., Hayenga, K.J., Kodama, K.H., Carlomagno, L.P., and Thompson, S.A. (1990). Molecular cloning and deletion of the aspergillopepsin A gene from *Aspergillus awamori*. *Gene 86*:153-162.

Boel, E., Hansen, M.T., Hjort, I., Hoegh, I., and Fiil, N.P. (1984). Two different types of intervening sequences in the glucoamylase gene from *Aspergillus niger*. *EMBO J. 3*:1581-1585.

Boel, E., Christensen, T., and Woldike, H.F. (1987). European Patent Application #0 238 023.

Buxton, F.P., Gwynne, D.I., and Davies, R.W. (1985). Transformation of *Aspergillus niger* using the *argB* gene of *Aspergillus nidulans*. *Gene 37*:207-214.

Chen, C.Y., Oppermann, H., and Hitzman, R.A. (1984). Homologous versus heterologous gene expression in the yeast, *Saccharomyces cerevisiae*. *Nucleic Acids Res. 12*:8951-8970.

Christensen, T., Woeldike, H., Boel, E., Mortensen, S.B., Hjortshoej, K., Thim, L., and Hansen, M.T. (1988). High level expression of recombinant genes in *Aspergillus oryzae*. *Bio/Technology 6*:1419-1422.

Cohen, B.L. (1977). The proteases of *Aspergilli*. In *The Genetics and Physiology of Aspergillus*. Smith, J.E., and Pateman, J.A., eds, Academic Press, London, pp. 281-292.

Cullen, D., Gray, G.L., Wilson, L.J., Hayenga, K.J., Lamsa, M.H., Rey, M.W., Norton, S., and Berka, R.M. (1987a). Controlled expression and secretion of bovine chymosin in *Aspergillus nidulans*. *Bio/Technology 5*:369-376.

Cullen, D., Leong, S.A., Wilson, L.J., and Henner, D.J. (1987b). Transformation of *Aspergillus nidulans* with the hygromycin-resistance gene, *hph*. *Gene 57*:21-26.

Emtage, J.S., Angal, S., Doel, M.T., Harris, T.J.R., Jenkins, B., Lilley, G., and Lowe, P.A. (1983). Synthesis of calf prochymosin (prorennin) in *Escherichia coli*. *Proc. Natl. Acad. Sci. USA 80*:3671-3675.

Finkelstein, D.B. (1987). Improvement of enzyme production in *Aspergillus*. *Antonie van Leeuwenhoek J. Microbiol. 53*:349-352.

Foltmann, B. (1979). Prochymosin and chymosin (prorennin and rennin). *Methods Enzymol. 19*:421-436.

Foltman, B., Pedersen, V.B., Jacobsen, H., Kauffman, D., and Wybrandt, G. (1977). The complete amino acid sequence of prochymosin. *Proc. Natl. Acad. Sci. USA 74*:2321-2324.

Foltman, B., Pedersen, V.B., Kauffman, D., and Wybrandt, G. (1979). The primary structure of calf chymosin. *J. Biol. Chem. 254*:8447-8456.

Goff, C.G., Moir, D.T., Kohno, T., Gravius, T.C., Smith, R.A., Yamasaki, E., and Taunton-Rigby, A. (1984). Expression of calf prochymosin in *Saccharomyces cerevisiae*. *Gene 27*:35-46.

Gray, G.L., Hayenga, K., Cullen, D., Wilson, L.J., and Norton, S. (1986). Primary structure of *Mucor miehei* aspartyl protease: evidence for a zymogen intermediate. *Gene 48*:41-53.

Gwynne, D.I., Buxton, F.P., Williams, S.A., Garven, S., and Davies, R.W. (1987). Genetically engineered secretion of active human interferon and a bacterial endoglucanase from *Aspergillus nidulans*. *Bio/Technology 5*:713-719.

Harkki, A., Uusitalo, J., Bailey, M., Penttilä, M., and Knowles, J.K.C. (1989). A novel fungal expression system: secretion of active calf chymosin from the filamentous fungus *Trichoderma reesei*. *Bio/Technology 7*:596-603.

Harris, T.J.R., Lowe, P.A., Lyons, A., Thomas, P.G., Eaton, M.A.W., Millican, T.A., Patel, T.P., Bose, C.C., Carey, N.H., and Doel, M.T. (1982). Molecular cloning and nucleotide sequence of cDNA coding for calf preprochymosin. *Nucleic Acids Res. 10*:2177-2187.

Kawaguchi, Y., Shimizu, N., Nishimori, K., Uozumi, T., and Beppu, T. (1984). Renaturation and activation of calf prochymosin produced in an insoluble form in *Escherichia coli*. *J. Biotechnol. 1*:307-315.

Kawaguchi, Y., Kosugi, S., Sasaki, K., Uozumi, T., and Beppu, T. (1987). Production of chymosin in *Escherichia coli* cells and its enzymatic properties. *Agric. Biol. Chem. 51*:1871-1877.

Kelly, J.M., and Hynes, M.J. (1985). Transformation of *Aspergillus niger* by the *amdS* gene of *Aspergillus nidulans. EMBO J. 4*:475-479.

Kolar, M., Punt, P. J., van den Hondel, C.A.M.J.J., and Schwab, H. (1987). Transformation of *Penicillium chrysogenum* using dominant selection markers and expression of an *Escherichia coli lacZ* fusion. In *Book of Abstracts, 19th Lunteren Lectures on Molecular Genetics, Molecular Genetics of Yeasts and Filamentous Fungi and its Impact on Biotechnology*, Lunteren, The Netherlands, 29 September-2 October 1987, p. 41.

Lamsa, M., and Bloebaum, P. (1989). Mutation and screening to increase chymosin yield in a genetically engineered strain of *Aspergillus awamori. J. Ind. Microbiol.*, in press.

Marston, F.A.O., Lowe, P.A., Doel, M.T., Schoemaker, J.M., White, S., and Angal, S. (1984). Purification of calf prochymosin (prorennin) synthesized in *Escherichia coli. Bio/Technology 2*:800-804.

McCaman, M.T., Andrews, W.H., and Files, J.G. (1985). Enzymatic properties and processing of bovine prochymosin synthesized in *Escherichia coli. J. Biotechnol. 2*:177-190.

Mellor, J., Dobson, M.J., Roberts, N.A., Tuite, M.F., Emtage, J.S., White, S., Lowe, P.A., Patel, T., Kingsman, A.J., and Kingsman, S.M. (1983). Efficient synthesis of enzymatically active calf chymosin in *Saccharomyces cerevisiae. Gene 24*:1-14.

Moir, D., Mao, J., Schumm, J.W., Vovis, G.F., Alford, B.L., and Taunton-Rigby, A. (1982). Molecular cloning and characterization of double-stranded cDNA coding for bovine chymosin. *Gene 19*:127-138.

Moir, D.T., Mao, J., Duncan, M.J., Smith, R.A., and Kohno, T. (1985). Production of calf chymosin by the yeast *S. cerevisiae*. In *Developments in Industrial Microbiology*, Vol. 26. Underkofler, L., ed., Society for Industrial Microbiology, p. 75-85.

Nishimori, K., Kawaguchi, Y., Hidaka, M., Uozumi, T., and Beppu, T. (1982). Nucleotide sequence of calf prorennin cDNA cloned in *Escherichia coli. J. Biochem. 91*:1085-1088.

Nishimori, K., Kawaguchi, Y., Hidaka, M., Uozumi, T., and Beppu, T. (1982b). Expression of cloned calf prochymosin gene sequence in *Escherichia coli. Gene 19*:337-344.

Nishimori, K., Shimizu, N., Kawaguchi, Y., Hidaka, M., Uozumi, T., and Beppu, T. (1984). Expression of cloned calf prochymosin cDNA under control of the tryptophan promoter. *Gene 29*:41-49.

Nunberg, J.H., Meade, J.H., Cole, G., Lawyer, F.C., McCabe, P., Schweickart, V., Tal, R., Wittman, V.P., Flatgaard, J.E., and Innis, M.A. (1984). Molecular cloning and characterization of the glucoamylase gene of *Aspergillus awamori. Mol. Cell. Biol. 4*:2306-2315.

Pedersen, V.B., Christensen, K.A., and Foltmann, B. (1979). Investigations on the activation of bovine prochymosin. *Eur. J. Biochem. 94*:573-580.

Punt, P.J., Oliver, R.P., Dingemanse, M.A., Pouwels, P.H., and van den Hondel, C.A.M.J.J. (1987). Transformation of *Aspergillus* based on the hygromycin B resistance marker from *Escherichia coli*. *Gene 56*:117-124.

Rajagopalan, T.G., Stein, W.H., and Moore, S. (1966). The inactivation of pepsin by diazoacetylnorleucine methyl ester. *J. Biol. Chem. 241*:4295-4297.

Revuelta, J.L., and Jayaram, M. (1986). Transformation of *Phycomyces blakesleeanus* to G-418 resistance by an autonomously replicating plasmid. *Proc. Natl. Acad. Sci. USA 83*:7344-7347.

Rowlands, R.T., and Turner, G. (1973). Nuclear and extranuclear inheritance of oligomycin resistance in *Aspergillus nidulans*. *Mol. Gen. Genet. 126*:201-216.

Smith, R.A., Duncan, M.J., and Moir, D.T. (1985). Heterologous protein secretion in yeast. *Science 229*:1219-1224.

Stepanov, V.A. (1985). Fungal aspartyl proteinases. In *Aspartic Proteinases and their Inhibitors*, Kostka, V., ed., deGruyter, Berlin, pp. 27-40.

Tang, J. (1979). Evolution in the structure and function of carboxyl proteases. *Mol. Cell. Biochem. 26*:93-109.

Tang, J.P., Sepulveda, J., Marciniszyn, J. Jr., Chen, K.C.S., Huang, W.Y., Tao, N., Liu, D., and Lanier, J.P. (1973). Amino acid sequence of porcine pepsin. *Proc. Natl. Acad. Sci. USA 70*:3437-3439.

Tonuochi, N., Shoun, H., Uozumi, T., and Beppu, T. (1986). Cloning and sequencing of a gene for *Mucor* rennin, an aspartate protease from *Mucor pusillus*. *Nucleic Acids Res. 14*:7557-7568.

Turnbull, I.F., Rand, K., Willetts, N.S., and Hynes, M.J. (1989). Expression of the *Escherichia coli* enterotoxin subunit B gene in *Aspergillus nidulans* directed by the *amdS* promoter. *Bio/Technology 7*:169-174.

Upshall, A., Kumar, A. A., Bailey, M.C., Parker, M.D., Favreau, M.A., Lewison, K.P., Joseph, M.L., Maraganore, J.M., and McKnight, G.L. (1987). Secretion of active human tissue plasminogen activator from the filamentous fungus *Aspergillus nidulans*. *Bio/Technology 5*:1301-1304.

van den Berg, J.A., van der Laken, K.J., van Ooyen, A.J.J., Renniers, T.C.H.M., Rietveld, K., Schaap, A., Brake, A.J., Bishop, R.J., Schultz, K., Moyer, D., Richman, M., and Shuster, J.R. (1990). *Kluyveromyces* as a host for heterologous gene expression: expression and secretion of prochymosin. *Bio/Technology 8*:135-139.

van Hartingsveldte, W., Mattern, I.E., van Zeijl, C.M.J., Pouwels, P.H., and van den Hondel, C.A.M.J.J. (1987). Development of a homologous transformation system for *Aspergillus niger* based on the *pyrG* gene. *Mol. Gen. Genet. 206*:71-75.

Viebrock, A., Perz, A., and Sebald, W. (1982). The imported preprotein of the proteolipid subunit of the mitochondrial ATP synthase from *Neurospora crassa*. Molecular cloning and sequencing of the mRNA. *EMBO J. 1*:565-571.

Ward, M. (1989a). Production of calf chymosin by *Aspergillus awamori*. In *Genetics and Molecular Biology of Industrial Microorganisms*, Hershberger, C. L., Queener, S.W., and Hegeman, G., eds., American Society for Microbiology, Washington, DC, pp. 288-294.

Ward, M. (1989b). Heterologous gene expression in *Aspergillus*. In *Proceedings of the EMBO-ALKO Workshop on Molecular Biology of Filamentous Fungi*, Nevalainen, H., and Penttilä, M., eds., Foundation for Biotechnical and Industrial Fermentation Research, Helsinki, *6*:119-128.

Ward, M., and Turner, G. (1986). The ATP synthase subunit 9 gene of *Aspergillus nidulans*: sequence and transcription. *Mol. Gen. Genet. 205*:331-338.

Ward, M., Wilkinson, B., and Turner, G. (1986). Transformation of *Aspergillus nidulans* with a cloned, oligomycin-resistant ATP synthase subunit 9 gene. *Mol. Gen. Genet. 202*:265-270.

Ward, M., Wilson, L.J., Carmona, C.L., and Turner, G. (1988). The *oliC3* gene of *Aspergillus niger*: isolation, sequence and use as a selectable marker for transformation. *Curr. Genet. 14*:37-42.

Ward, M., Wilson, L.J., Kodama, K.H., Rey, M.W., and Berka, R.M. (1990). Improved production of chymosin in *Aspergillus* by expression as a glucoamylase-chymosin fusion. *Bio/Technology*, in press.

5
Fungal Aspartic Proteinases

Sheryl A. Thompson

Genencor International, South San Francisco, California

I. INTRODUCTION

Aspartic proteinases are a distinct group of endoproteolytic enzymes that are found widely throughout eukaryotic systems from fungi to higher plants and mammals. Mammalian aspartic proteinases have been isolated from such organs as stomach (chymosin, gastricin, pepsin), kidney (renin), and from cellular compartments such as lysozomes (cathepsin D) (Tang and Wong, 1987). In the vacuoles of the yeast *Saccharomyces cerevisiae*, an aspartic proteinase is produced that is involved in the processing of vacuolar proteins to their mature forms (Ammerer et al., 1986). Plants such as squash and cucumber also produce aspartic proteinases. Recently, several retroviruses, including HIV isolates (implicated as the causative agents of AIDS), have been found to code for an aspartic proteinase that is essential for proteolytic processing of viral polypeptides and viral infectivity (Hansen et al., 1988). Many strains of filamentous fungi secrete aspartic proteinases, and a number of these enzymes have been used commercially. Fungi from such diverse genera as *Mucor* (Bech and Foltmann, 1981), *Aspergillus* (Ostoslavskaya et al., 1986), *Endothia* (Barkholt, 1987),

107

Rhizopus (Grahm et al., 1973), and *Penicillium* (Sodek and Hoffman, 1970) produce aspartic proteinases, which have been purified and studied. This chapter will focus on studies of fungal aspartic proteinases, with special emphasis given to structure-function relationships and posttranslational modifications that are currently being investigated by molecular biological methods.

II. INDUSTRIAL USES FOR ASPARTIC PROTEINASES

Perhaps the most well-known industrial application of fungal aspartic proteinases involves their use in the dairy industry as milk-coagulating enzymes. This application was brought about because of periodic shortages in the availability of calf chymosin, which is the preferred milk-coagulating enzyme for cheese manufacture. Proteinases from *Mucor miehei, M. pusillus, Irpex lacteus,* and *Endothia parasitica* have been used as chymosin substitutes. Their activities can be controlled to give a high degree of milk coagulation with a minimum of continued proteolysis, which could result in an off-flavored cheese. The proteolytic activities of some fungal aspartic proteinases can actually be used to develop the flavor of certain cheeses. For example, Genencor has used a blend of proteinase and lipase from *Aspergillus oryzae* to assist the ripening of cheddar cheeses (Arbige et al., 1986). The aspartic proteinase from *A. oryzae* has also been used in the commerical baking industry to improve dough texture and to contribute to the flavor and crust color of certain breads (Genencor). A food-grade aspartic proteinase from *A. niger* has also been used to treat the proteinaceous haze that develops in certain white wines (Bakalinsky and Boulton, 1985). These protein-unstable wines are considered defective and unmarketable unless the protein precipitate is removed.

III. BIOCHEMICAL AND STRUCTURAL PROPERTIES
OF ASPARTIC PROTEINASES

Characteristically, aspartic proteinases function at low or neutral pH, hence the initial name of *acid protease* was derived. The activity of an individual aspartic proteinase is defined by its interaction with specific protease inhibitors. Pepstatin A is an N-acylpentapeptide (isovaleryl-Val-Val-Sta-Ala-Sta, where *Sta* represents statine, 4-amino-3-hydroxy-6-methyl-heptanoic acid) isolated from *Streptomyces lignacolum*. Pepstatin acts as a substrate analogue by binding tightly to aspartic proteinases in a 1:1 molar ratio (Umezawa et al., 1970). Two active-site affinity labels, diazoacetyl norleucine methyl ester (DAN) and 1,2-epoxy-3-(p-nitrophenoxy) propane (EPNP), are also used to characterize the catalytic residues of aspartic proteinases (Rajogopalan et al., 1966). These affinity labels react with the side-chain carboxyl group of aspartic acid residues 215 and 32 (porcine pepsin numbering system), respectively, rendering the enzyme inactive. These data indicate that aspartic acid residues 215 and 32 function in the catalytic mechanism; hence, the concise name of *aspartic proteinase* was derived.

As a group, fungal aspartic proteinases are highly homologous, regardless of the source. Analysis of the primary structure of several aspartic proteinases reveals areas of strong amino acid sequence homology, especially in and around the active-site residues (Figure 1). Comparison of the tertiary structures also shows strong similarities between species. The three-dimensional structures of penicillopepsin (Hsu et al., 1977), endothiapepsin (Blundell et al., 1985), rhizopuspepsin (Bott et al., 1982), and porcine pepsin (Andreeva et al., 1985) have been determined by x-ray crystallography. Generally, each molecule consists of approximately 320-360 amino acids. These are arranged into a secondary structure consisting primarily of an 18- to 20-strand β-pleated sheet structure that twists in an anticlockwise direction, forming a globular-shaped protein with two symmetrical lobes (Figure 2). Residing between these two lobes is an extended binding cleft, capable of accomodating a substrate of approximately 7-8 amino acids, as predicted by the binding specificity of both protein and synthetic peptide substrates (James et al., 1982). Near the center of this pocket lie the two active aspartic acid residues 215 and 32. Each lobe contains one of the aspartic acid residues, which protrudes into the substrate binding cleft.

Studies of aspartic protease three-dimensional structures reveal a two-fold axis of symmetry between the two lobes of the enzyme. The axis of symmetry passes through the active center of the molecule, separating the two active aspartic acid residues. In pepsin, 14 identical amino acids exist in the sequence alignment between the two lobes. There are, however, 61 amino acids of the protein sequence that align topologically with each other when comparing the crystal structures of the two lobes (Blundell et al., 1979). Thus, even though the protein sequences of the two lobes are not identical, structurally they produce very similar units. A second axis of symmetry has been found to exist within each of the two lobes of the protein. Blundell and co-workers (1979) found three-dimensional structural repeats within the two lobes of endothiapepsin. Each lobe of the protein was found to be composed of two sets of four twisted antiparallel

```
Porcine pepsin (PEP)        I G D E - P L E N Y L D T E - Y F
Bovine Chymosin (CHY)       G E V A S V - P L T N Y L D S Q - Y F

Aspergillopepsin (AAP)      S K G S A V T T P Q N - - N D - E E Y L
Penicillopepsin (PJP)       A A S G V A T N T P T A - - N D - E E Y I
Endothiapepsin (EPP)        S T G S A T T T P I D S L - D - D A Y I
R.chinensis (RCP)           A G V G T V P M T D Y G - - N D V E Y Y -
R. niveus (RNP)             A   S G S V P M V D Y E - - N D V E Y Y -
M. miehei (MMP)             A A A D G S V D - T P G Y Y D F D L E E Y A
M. pusillus (MPP)           A E G D G S V D - T P G L Y D F D L E E Y A
Highly conserved            .     * . .     * *             *   * . *


                                    32      35
(PEP)       G T I G I G T P A Q D F T V I F D T G S S N L W V P S V Y C
(CHY)       G K I Y L G T P P Q E F T V L F D T G S S D F W V P S I Y C

(AAP)       T P V T V G K S T L H L D - - F D T G S A D L W V F S D E L
(PJP)       T P V T I G G T T L N L N - - F D T G S A D L W V F S T E L
(EPP)       T P V Q I G T P A Q T L N L D F D T G S S D L W V F S S E T
(RCP)       G Q V T I G T P G K K F N L D F D T G S S D L W I A S T L C
(RNP)       G E V T V G T P G I K L K L D F D T G S S D M W F - S T L C
(MMP)       I P V S I G T P G Q D F L L L F D T G S S D T W V P H K G C
(MPP)       I P V S I G T P G Q D F Y L L F D T G S S D T W V P H K G C
            . * * . * * . * . .       . .   * * * * * . * . * * . . . .

(PEP)       - S S L A C - S D H N Q F N P D D S S T F E A T S Q E L S I
(CHY)       - K S N A C - K N H Q R F D P R K S S T F Q N L G K P L S I

(AAP)       P S S E - - - - - G H D L Y T P S S S S T K L S G Y T W D I
(PJP)       P A S Q - - - Q S G H S V Y N P S A T G K E L S G Y T W S I
(EPP)       S E V D - - - G Q T I Y T P S K S T T A K L L S G A T W S I
(RCP)       T N - - - C G S R Q T K Y D P K Q S S T Y Q A D G R T W S I
(RNP)       - S S -   C S N S H T K Y D P K K S S T Y A A D G R T W S I
(MMP)       T K S E G C V G S R - F F D P S A S S T F K A T N Y N L N I
(MPP)       D N S E G C V G K R - F F D P S S S S T F K E T D Y N L N I
                                                                    *

(PEP)       T Y G T G S M - - T G I L G Y D T V Q V G G I S D T N Q I F
(CHY)       H Y G T G S M - - Q G I L G Y D T V T V S N I V D I Q Q T V

(AAP)       S Y G D G S S A A S G D V Y R D T V T V G G V T T N K V E A
(PJP)       S Y G D G S S - A S G N V F T D S V T V G G V T A H A V Q A
(EPP)       S Y G D G S S - S S G D V T D T V S V G G V T G Q A V E S
(RCP)       S Y G D G S S - A S G I L A K D N V M L G G L L G Q T I Q L
(RNP)       S Y G D G S S - A S G I L A T D N V N L G G L L I K K Q T I
(MMP)       T Y G T G G A D G L Y F E D S I A I G D I T V T K Q I L A Y
(MPP)       T Y G T G G A N G I Y F R T S I T V G G A T V K Q Q T L A Y
            . * * . * . .   . . .         . * .       . .

(PEP)       G S L E T E P G S F L Y Y A P F - - - D G I L G L A Y P S I
(CHY)       G L S T Q E P G D V F T Y A E F - - - D G I L G M A Y P S L

(AAP)       - - A S K I S S E F V Q N T A N - - - D G L L G L A F S S I
(PJP)       - - A Q Q I S A Q F Q Q D T N N - - - D G L L G L A F S S I
(EPP)       - - A K K V S S S F T E D S T I - - - D G L L G L A F S T L
(RCP)       - - A K R Q A A S F A N G P - N - - - D G L L G L G F D T I
(RNP)       E L A L R E S S A F A T D V I - - - - D G L L G L G F N T I
(MMP)       V D N V R G P T A E Q S P N A I D F L D G I F G A A Y P D N
(MPP)       V D N V S G P T A E Q S P D S E L F L D G I F G A A Y P D N
            .       . . .       . * * * . * . . . .

(PEP)       S A S G A T P V F D N L W D Q G L V S Q D L F S V Y L S S N
(CHY)       A S E Y S I P V F D N M M N R H L V A Q D L F S V Y M D R D

(AAP)       N T V Q P K A Q T T F F D T V K S Q L D L F A V Q L K H D A
(PJP)       N T V Q P Q S Q T T F F D T V K S S L P L F A V A L K H Q Q
(EPP)       N T V S P T Q Q K T F F D N A S L D S P V F T A D L G Y H A
(RCP)       T T V R G V K T P N D N L I S Q G L R P I F G V Y L G K A S
(RNP)       T T V R G V K T P V D N L I S Q G L I S - - - Q G L I S R P
(MMP)       T A M E A E Y G S T Y N - T V H V N - - L Y K Q G L I S S P
(MPP)       T A M E A E Y G D T Y N - T V H V N - - L Y K Q G L I S S P
            . . .           . * . . . . . .       . .       *

(PEP)       D D S G S V V L L G G I D S S Y Y T G S L N W V P V S V E -
(CHY)       G Q E S M - L T L G A I D P S Y Y T G S L H W V P V T V Q Q

(AAP)       - P G - - V Y D F G Y I D D S K Y T G S I T Y T D A D S S Q
(PJP)       - P G - - V Y D F G F I D S S K Y T G S L T Y T G V D N S Q
(EPP)       - P G - - T Y N F G F I D T T A Y T G S I T Y T A V S T K Q
(RCP)       N G G G G E Y I F G G Y D S T K F K G S L T T V P I D N S R
(RNP)       I F G V Y L G K Q S N G G G G E Y I F G G Y D S S N F K G S
(MMP)       L F S - - V Y M N T N S G T G E V - F G G V N N T L L S G D
(MPP)       V F S - - V Y M N T D G G G Q V V F G G V N N T L L G G D
            . . .         * . . .   . .       . .

(PEP)       - - - - - - - - - - - - - G Y W Q I T L D S I T M D G E T I
(CHY)       - - - - - - - - - - - - - Y W Q F T V D S V T I S G V V V

(AAP)       - - - - - - - - - - - - - G Y W G F S T D G Y S D G S S S S
(PJP)       - - - - - - - - - - - - - G F W S F N V D S Y A G S Q S G D
(EPP)       - - - - - - - - - - - - - G F W E W T G Y A V G S G T F K S
(RCP)       - - - - - - - - - - - - - G W W G I T R A T V G T S T V A S
(RNP)       L T T V P I D N S E - - - G F W G V T V N S T K I G G T T S
(MMP)       I A Y T D V M S R Y G G Y Y F W D A P V T G I T V D G S A A
(MPP)       I Q Y T D V L K S R G G Y F F W D A P V T G V K I D G S D A
                          . . *         . .     . . .

                                    215
(PEP)       A C - S G G C Q A I V D T G T S L L T G P T S A I A I N Q S
(CHY)       A C - E G G C Q A I L D T G T S K L V G P S S D I L N I Q Q
```

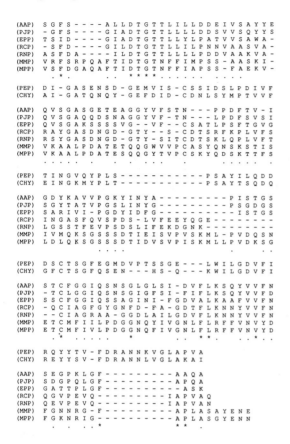

Figure 1 Amino acid sequence homologies of fungal aspartic proteinases and mammalian gastric proteinases.

strands of β sheet. Therefore, it is possible to postulate that a primoridial aspartic proteinase might have existed that was one-quarter the size of the current protein (i.e., about 90 amino acids). This protein would have been composed of an ancestral β-sheet unit of approximately 45 residues. The gene encoding this polypeptide may have undergone a duplication event followed by a gene fusion, resulting in a protein similar to one lobe of the aspartic proteinases known today. A second duplication event followed by a gene fusion could then have occurred, resulting in a bilobal molecule with fourfold structural repeats.

The strong similarities seen within the fungal aspartic proteinases can also be seen when comparing fungal and mammalian proteinases. Comparisons in the primary protein sequences of porcine pepsin and bovine chymosin with several fungal aspartic proteinases are shown in Figure 1.

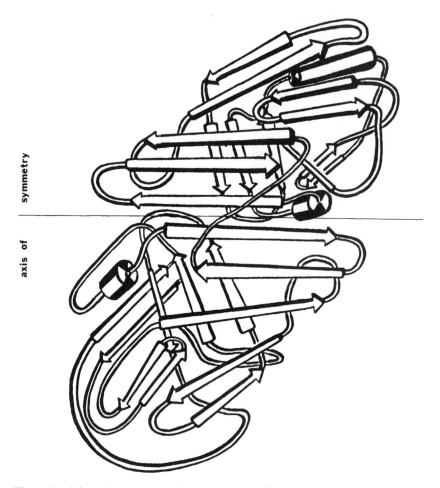

Figure 2 Schematic representation of the tertiary structure of fungal aspartic proteinases showing the twofold axis of symmetry (Tang, 1979).

The alignment of porcine pepsin and aspergillopepsin A from *A. awamori* reveals a 32% homology between the two proteins, while chymosin and aspergillopepsin A are 28% homologous. Comparisons of the x-ray crystal structure of porcine pepsin (Andreeva et al., 1985) with those of the fungal aspartic proteinases reveals a high degree of structural homology, suggesting that aspartic proteinases from widely divergent species may have evolved from a common ancestral gene. Hence, a primordial aspartic proteinase may have existed long before the divergence of mammals took place (Tang, 1979). It was once thought that fungal aspartic proteinases were the closest example of what the primordial aspartic proteinase was like, however, the primary sequences of retroviral aspartic proteinases (Hansen et al., 1988) suggests that they may, in fact, represent the most primitive form of aspartic proteinases.

Retroviral proteases contain the conserved amino acid sequence Asp-Thr-Gly, which is also conserved in the active site of other aspartyl proteases. Analysis of the aspartyl protease from the human immunodeficiency virus HIV revealed that this conserved sequence could be similar to that which is contained in a single domain of the fungal aspartyl proteases. Pearl and Taylor (1987) postulated that the retroviral proteases in active form would be a dimer. The crystal structure of the HIV protease has been determined (Miller et al., 1989; Navia et al., 1989). It shows that the HIV protease is functionally active as a dimer and thus forms a bi-lobal molecule with an active-site pocket containing two aspartic acid residues. This data supports the theory that aspartic proteases once existed as a smaller molecule, which then underwent a gene duplication event to form the present-day aspartic proteases.

IV. ZYMOGEN STRUCTURE OF FUNGAL ASPARTIC PROTEINASES DEDUCED FROM DNA SEQUENCE DATA

Secreted aspartic proteinases from both fungal and mammalian cells contain amino-terminal extensions of approximately 60-70 amino acids that are not found in the mature forms of the enzymes. These amino-terminal extensions are composed of two regions: (a) a signal peptide of 15-20 amino acids, which aids in the secretion of the enzyme (for a review of signal peptides, see Perlman and Halvorson, 1983) and (b) a 40-50 amino acid propeptide, which is thought to stabilize the enzyme in an inactive conformation (i.e., zymogen or proenzyme) until it is "activated" by proteolytic processing (Figure 3). A signal peptide typically contains a core sequence of 10 or more consecutive hydrophobic amino acids, whereas the propeptide portion of an aspartic proteinase is relatively rich in positively charged residues such as lysine, arginine, and histidine. No zymogen precursors of aspartic proteinases have yet been isolated from filamentous fungi. Evidence for their existence, however, has been derived from DNA sequence analysis of cloned aspartyl protease genes. The aspartic proteinase

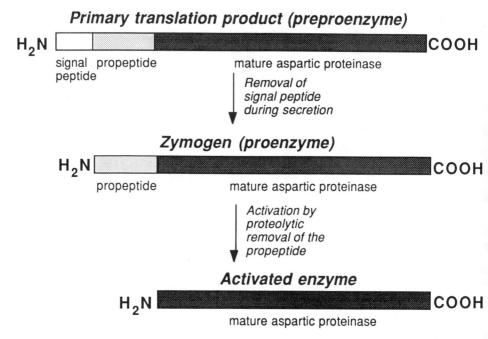

Figure 3   Posttranslational processing of fungal aspartic proteinases. The signal peptide is removed during secretion into the lumen of the endoplasmic reticulum. The propeptide portion is subsequently removed by either autocatalytic processing or by proteolytic activation by another enzyme.

genes from *A. awamori, Mucor miehei, M. pusillus, Rhizopus niveus,* and a partial cDNA clone from *R. chinensis* have been isolated and studied.

A partial cDNA clone of rhizopuspepsin, pl 5, was isolated by Delaney et al. (1987) from the filamentous fungus *R. chinensis.* Nucleotide sequence analysis of the 1140-bp cDNA segment revealed a possible propeptide at the amino terminus of rhizopuspepsin. This propeptide fragment was found to be rich in positively charged amino acids: five lysines and one histidine in the 26 residues that were deduced. The authors concluded that the cDNA clone was incomplete at the 5′ end and thus lacking a signal peptide because (a) no methionine residue was found in the deduced amino acid sequence, and (b) Northern blotting experiments performed with *R. chinensis* RNA probed with the cDNA clone revealed a single 1.3-kb mRNA species, while the cloned cDNA was only 1.14 kb. Thus, the cDNA clone isolated lacked approximately 160 bp of the 5′ end. A more complete clone of the 5′ end of the gene has subsequently been isolated (J. Tang, personal communication). Although the N-terminal methionine is still not present on the clone, a potential hydrophobic signal sequence can be seen upstream of the potential propeptide. Nucleotide sequence analysis of the complete clone will be required to determine the exact length of the signal peptide.

Kobayashi et al. (1989) isolated a partial cDNA clone of the aspartic proteinase from *Irpex lacteus*. *I. lactus* produces an aspartyl protease that is a good substitute for chymosin in producing fiber-structured cheeses (i.e., gouda and cheddar). A cDNA library was probed with a synthetic oligonucleotide corresponding to the N terminus of the protein. The isolated 1.2-kb cDNA clone contained codons for the mature amino acid sequence but no propeptide or signal-peptide sequences. Northern blots probed with the cloned gene revealed a message of 1.5 kb, thus the clone is missing approximately 300 bp.

A genomic clone of the aspartic proteinase gene from *M. miehei* was isolated by Gray et al. (1986) using degenerate oligonucleotide probes corresponding to the known protein sequence. The gene is contained within a 2.5-kb *Cla*I-*Bam*HI restriction fragment and comprises an open reading frame of 1290 bp with no introns. The putative start codon was located 207 bp upstream of the first codon for the mature proteinase, suggesting that the enzyme is synthesized as a zymogen with a 69 amino acid preproregion. The first 22 amino acids of this region contained core of 10 hydrophobic residues, typical of signal peptides (Perlman and Halvorson, 1983). The core region ends with a serine residue and the proposed signal-peptide cleavage site of Thr-Thr-Ala. The remaining 47 amino acids that precede the amino terminus of the mature enzyme probably represent the propeptide. This propeptide region contains several positively charged amino acid residues (two arginine and seven lysine).

The aspartic proteinase gene from *M. pusillus* was isolated from a cosmid bank made with genomic DNA (Tonouchi et al., 1986). Nucleotide sequence analysis of the clone revealed that the mature proteinase from *M. pusillus* is 83% homologous at the amino acid level with the enzyme from *M. miehei*. This homology extended into the signal peptide and propeptide regions as well, but some minor differences were observed. Like the *M. miehei* proteinase gene, the *M. pusillus* clone contained an open reading frame without introns, encoding a mature enzyme of 361 amino acids. The prepro region of the *M. pusillus* aspartic proteinase (66 amino acids) was shorter than the corresponding region of the *M. miehei* proteinase. This difference is reflected mainly in the length of the proposed signal peptides. While the signal peptides of both proteinases contain hydrophobic core regions of 10 amino acids ending with a Ser residue, the entire signal peptide for *M. pusillus* proteinase is 18 amino acids long, compared with 22 amino acids for *M. miehei*. The proposed signal-peptide cleaveage site for the *M. pusillus* enzyme, Ser-Phe-Ala, is also different than that found in *M. miehei*. The propeptide region of the *M. pusillus* aspartic proteinase, comprised of 48 amino acids, is similar to that of *M. miehei* and is highly charged.

A genomic DNA segment encoding the aspergillopepsin A gene from *A. awamori* was isolated on a 9-kb *Eco*RI restriction fragment (Berka et al., 1990). Unlike the *Mucor* genes, the aspergillopepsin A gene was found to contain three small introns. DNA sequence data showed that the gene is encoded by four exons of 320, 278, 249, and 338 bp, separated by three introns of 51, 52, and 59 bp. A putative signal peptide and propeptide was found to be present that was of similar size and composition to those found in *Mucor* species. The proposed signal peptide for aspergillopepsin A comprises the first 20 residues of this 69 amino acid sequence, including a hydrophobic core of eight residues and ending with a potential cleavage site of Val-Ser-Ala. The next 49 amino acids represent the putative propeptide region, which is rich in basic residues (four arginine and four lysine). The mature aspergillopepsin protein is somewhat shorter than the *Mucor* enzymes: 325 amino acids in aspergillopepsin A versus 361 amino acids for *Mucor* species.

Horiuchi et al. (1988) isolated the aspartyl protease gene from *R. niveus* using degenerate oligonucleotide probes complementary to the primary protein sequence. DNA sequence analysis of the cloned gene revealed that a single intron of 64 bp was present, separating the gene into two exons of 326 and 845 bp. The placement of this intron near the 5' end of the gene is similar to the placement of the first intron present in aspergillopepsin A. In *R. niveus*, the intron is situated between amino acids 42 and 43, while the first intron of aspergillopepsin A resides between the second and third bases that encode amino acid 39. The proposed signal-propeptide portion of the *R. niveus* aspartyl protease is 66 amino acids in length. The signal peptide contained a potential core sequence of 11 amino acids and a possible cleavage site of Val-Glu-Ala. Comparison of the last 27 amino acids of the *R. niveus* propeptide with that of the incomplete propeptide sequence from *R. chinensis* shows a 54% homology between the two.

Comparisons of signal and propeptide regions of various fungal aspartic proteinases can be seen in Figure 4. While the primary amino acid sequence homology between the mature enzymes from *M. miehei* and *A.*

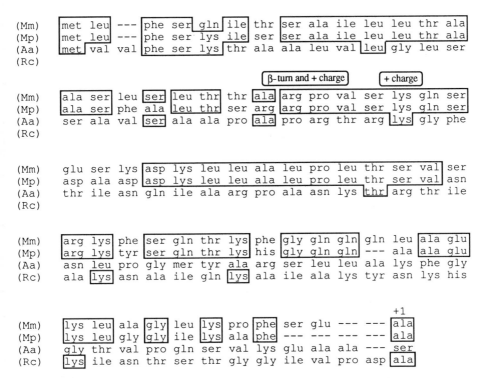

Figure 4 Amino acid sequence alignment of the signal peptide and propeptide portions of the aspartic proteinases from *Aspergillus awamori* (Aa), *Mucor miehei* (Mm), *Mucor pusillus* (Mp), and *Rhizopus chinensis* (Rc).

awamori is 29%, the homology between their respective signal peptides and propeptides is only 20%. Figure 5 shows hydrophobicity plots for signal peptides and propeptides of the aspartic proteinases from *A. awamori*, *M. miehei*, *M. pusillus*, and *R. niveus*. Surprisingly, although the primary amino acid sequences are not homologous, similar hydrophobicity patterns are seen for all four, especially within the signal peptides. The hydrophobicity plots for the first 20 amino acids of the signal peptide are nearly identical for all four. The hydrophobicity patterns for the propeptides of these three fungal proteinases are not as similar as those for the signal peptides. The significance of these differences is unknown. In the alignment of aspartic proteinase propeptides of the zygomycete genera *Mucor* and *Rhizopus*, certain charged residues are highly conserved (Figure 4).

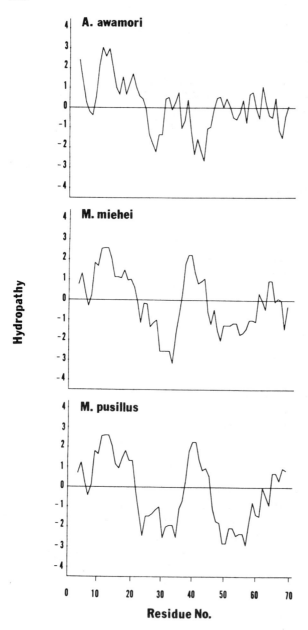

Figure 5 Hydropathy plots of the signal peptide and propeptide portions of several fungal aspartic proteinases. Hydrophobic areas are indicated by positive values, while hydrophilic regions are given negative values (Kyte and Doolittle, 1982).

It is tempting to speculate that these residues may be important for the proper interaction of the propeptide with the substrate binding cleft and active-site residues of the aspartic proteinases. These interactions may help to hold the molecule in an inactive conformation, as well as position the propeptide for proper cleavage and subsequent enzyme activation.

From the published DNA sequences it can be seen (Table 1) that the codon biases for the aspartic proteinase genes of *A. awamori, M. miehei, M. pusillus, R. niveus,* and *R. chinensis* are quite similar. As with many

Table 1 Comparison of the Codon Frequency of the Aspartyl Protease Genes from *Aspergillus awamori* (AP), *Rhizopus cheninsis* (RP), *Mucor miehei* (MM), and *Mucor pusillus* (MP).

Amino acid	Codons	AP	RP	MM	MP
Phe	UUU	5	7	12	7
	UUC	14	10	14	22
Leu	UUA	0	3	1	0
	UUG	2	7	6	12
	CUU	2	4	12	10
	CUC	7	8	8	6
	CUA	0	0	2	1
	CUG	15	0	4	2
Ile	AUU	3	8	11	7
	AUC	14	17	13	12
	AUA	0	0	0	0
Met	AUG	2	2	8	5
Val	GUU	9	9	10	13
	GUC	14	14	16	14
	GUA	0	0	1	3
	GUG	7	0	2	2
Ser	UCU	7	12	10	14
	UCC	17	8	12	14
	UCA	1	0	2	0
	UCG	5	0	10	3
	AGU	2	2	2	2
	AGC	21	8	9	9
Pro	CCU	6	8	8	9
	CCC	7	4	6	6
	CCA	2	1	5	4
	CCG	2	0	2	0

Table 1 (Continues)

Table 1 (Continued)

Amino acid	Codons	AP	RP	MM	MP
Thr	ACU	9	17	18	11
	ACC	21	11	11	19
	ACA	1	1	2	2
	ACG	6	2	4	0
Ala	GCU	8	14	16	13
	GCC	21	12	7	9
	GCA	2	0	10	7
	GCG	4	0	3	1
Tyr	UAU	0	7	7	2
	UAC	19	8	13	17
Stop	UAA	0	1	1	1
	UAG	1	0	0	0
	UGA	0	0	0	0
His	CAU	1	0	0	0
	CAC	2	1	2	3
Gln	CAA	0	12	10	7
	CAG	15	1	7	10
Asn	AAU	2	6	2	0
	AAC	11	15	20	20
Lys	AAA	1	2	3	7
	AAG	17	16	14	15
Asp	GAU	8	13	9	18
	GAC	19	11	16	16
Glu	GAA	3	6	6	5
	GAG	9	1	10	9
Cys	UGU	0	4	1	0
	UGC	2	0	3	4
Trp	UGG	3	4	3	2
Arg	CGU	1	6	2	1
	CGC	2	2	4	3
	CGA	0	0	1	2
	CGG	2	0	0	0
	AGA	0	1	1	0
	AGG	0	0	1	0
Gly	GGU	15	35	16	14
	GGC	19	10	18	23
	GGA	16	2	4	8
	GGG	0	0	0	0

other fungal genes, codons with a pyrimidine in the wobble position are preferred. Where codon choices offer only a purine in the third position, G is preferred over A. The arginine codons AGA and AGG are rarely used.

V. HETEROLOGOUS EXPRESSION OF ASPARTIC PROTEINASES

Many strains of filamentous fungi are able to secrete large amounts of protein in submerged culture. For example, strains of *A. awamori* are known to secrete more than 10 g/l of glucoamylase into the culture medium (see Chapter 4), and *T. reesei* strains have been reported to produce nearly 40 g/l of exo-cellobiohydrolase I (Nevalainen et al., 1988). This characteristic makes fungal hosts attractive for the production of heterologous proteins. Transformation systems for introducing foreign DNA have been described for several fungal species (see Chapter 1). In addition, fungi appear to be rather permissive in expressing genes from other fungi. For example, the *A. nidulans argB* and *amdS* genes can be used as selectable markers for the transformation of several other species (Buxton et al., 1985; Kelly and Hynes, 1985). The cloned aspartic proteinase genes from *M. miehei* and *M. pusillus* have been expressed in several filamentous fungal strains, as well as in *E. coli* and *S. cerevisiae* (Tonouchi et al., 1986; Gray et al., 1986). The transformed fungi were capable of expressing the heterologous proteinase gene and secreting the enzyme into the culture medium. Conversely, yeast and *E. coli* were less productive in the expression and secretion of the fungal proteinases.

Gray et al. (1986) constructed an expression vector for *E. coli* by placing the *M. miehei* aspartic proteinase gene under the transcriptional control of the *E. coli trp* promoter. Transformants obtained with this vector were fractionated into cytoplasmic/membrane and periplasmic fractions. Equal amounts of proteinase activity were found in the cytoplasmic/membrane and periplasmic fractions, however, the amount of proteinase detected by immunological blotting was much higher in the cytoplasmic/membrane fraction. Thus, most of the proteinase that was capable of entering the periplasmic space had been processed to such an extent that it was active. However, the majority of the *Mucor* proteinase that was produced by *E. coli* remained unprocessed within the cell.

A second expression vector for the transformation of *A. nidulans* was constructed by Gray et al. (1986) utilizing the *M. miehei* aspartic proteinase gene. *A. nidulans* appears to be a good host for studying heterologous aspartyl protease production, because it produces little if any aspartyl protease of its own. The entire aspartic proteinase gene of *M. miehei* plus 600 bp of 5′-flanking DNA was inserted into the transformation vector pDJB1 (Ballance and Turner, 1985), and the resulting plasmid was used to transform a *pyrG* auxotroph of *A. nidulans*. Hybridization analysis of cellular DNA isolated from the transformants indicated that, in most cases, several copies of the *Mucor* proteinase gene had integrated into the *A. nidulans* genome at random locations. Analysis of culture filtrates from the transformants revealed that the aspartic proteinase was being produced and secreted in an active form. No immunoreactive proteinase could be detected intracellularly in the transformants. These data suggest that the transcriptional, translational, and secretory signals of the *Mucor* aspartic proteinase gene are recognized and used by *A. nidulans*. Determination of the specific activity of the heterologous *M. miehei* aspartyl protease produced in *A. nidulans* revealed that it was similar to that of the wild-type enzyme. SDS-polyacrylamide gel electrophoresis revealed that the molecular weight of the *Mucor* gene product produced by *A. nidulans* appeared to be greater than that of the native enzyme produced by *M. miehei*. This phenomenon was apparently due to overglycosylation of the enzyme in *A. nidulans*. When both the native *Mucor* enzyme and the recombinant enzyme from *A. nidulans* are treated with endoglycosidase H, their mobilities on SDS-polyacrylamide gels are the same.

Christensen et al. (1987) have studied the expression of *M. miehei* aspartic proteinase in *A. oryzae*. A cDNA clone encoding the aspartic proteinase was isolated and placed in a vector that contained the *A. oryzae* α-amylase promoter and the *A. niger* glucoamylase terminator. This expression vector was introduced into *A. oryzae* by cotransformation with a second plasmid, which contained the selectable *amdS* gene from *A. nidulans*. Transformants were found to secrete active *Mucor* proteinase into the culture medium with yields ranging from 0.5 to 3.3 g/l, while no protease was detected in the untransformed strains. Characterization of the heterologous gene product showed that it was also hyperglycosylated compared to the native enzyme from *M. miehei*, while the specific activity was unchanged.

Dickinson et al. (1987) have expressed the *M. miehei* aspartic proteinase in *M. circinelloides* using the transcription, translation, and secretory components of the *M. miehei* gene. *M. circinelloides* is a mesophillic filamentous fungus and thus is separated from the thermophillic *Rhizomucor* species of *M. miehei*, *M. pusillus*, and *M. testacus*. Because of this difference, the native *M. circinelloides* aspartyl protease can be distinguished from that of the heterologous *M. miehei* gene product by two criteria:

(a) Each species has a different molecular weight, 34,000 Dal for *M. circinelloides* and 42,000 Dal for *M. miehei*, and (b) antibodies raised to each protease do not cross-react with the other. Thirty from a total of 40 *M. circenelloides* transformants secreted the heterologous gene product as demonstrated by rocket immunoelectrophoresis. The levels of extracellular proteinase (determined immunologically) ranged 1-12 mg/l. The excreted proteinase was enzymatically active and had the same apparent molecular weight as the aspartic proteinase from *M. miehei*. Thus, the enzyme appears to be glycosylated to a similar degree as that of the native *M. miehei* proteinase.

Yamashita et al. (1987) studied the expression of the *M. pusillus* aspartic proteinase in *S. cerevisiae*. The aspartic proteinase coding region, as well as the 5′ and 3′ noncoding regions of DNA, were placed into the autonomously replicating yeast vector pSS21. Supernatants and cell extracts from the resulting yeast transformants contained no enzymatic activity nor immunodetectable material. The authors concluded that the *M. pusillus* promoter signals were not functional in *S. cerevisiae*. A second expression vector was then constructed using the yeast *GAL7* promoter. Transformants derived with this vector secreted active proteinase into the culture medium when galactose was used as the carbon source, suggesting *GAL7*-dependent transcription of the heterologous gene. The signal peptide and propeptide regions of the *M. pusillus* proteinase were processed correctly, as revealed by amino acid sequencing of the proteinase purified from yeast culture supernatants. The yield of recombinant enzyme in the culture medium was more than 150 mg/l, but some inactive intracellular aspartic proteinase was also found. The enzyme secreted from yeast was also glycosylated differently than the native *Mucor* enzyme, yet the enzymatic properties of the recombinant enzyme remained nearly identical to those of the native enzyme.

In general, we can see that when filamentous fungi are transformed with heterologous aspartyl protease genes, the introduced promoter and secretory sequences are recognized by the fungus and are utilized to express the gene of interest. Propeptides and signal peptides are also recognized by the fungus and are processed in order to produce active enzymes, which are secreted into the culture medium. It is not yet known, however, if these processing events arise by the same mechanism as those which take place in the native host. Heterologous expression of aspartyl proteases between species often results in the hyperglycosylation of the mature enzyme. Why this phenomenon occurs is unknown. It is possible that the heterologous gene products pass through compartments of the secretory pathway (e.g., Golgi apparatus) less efficiently than native proteins and thus are made available to transglycosylases for a longer period of time. This hyperglycosylation does not, however, appear to affect the specific activity of the enzymes.

VI. RECOMBINANT SYSTEMS TO STUDY STRUCTURE-FUNCTION RELATIONSHIPS AND ACTIVATION OF ASPARTIC PROTEINASES

Activation of aspartic proteinases occurs by proteolytic removal of the propeptide, which is typically rich in positively charged amino acid residues such as lysine, arginine, and histidine (Figure 3). Although there is little known regarding the activation of fungal aspartic proteinases, the molecular mechanism for the activation of the corresponding mammalian enzymes has been studied in detail (Foltmann, 1981). Gastric proteinases apparently require only hydrogen ions in order to initiate limited activation. This observation may reflect a pH-dependent change in the conformation of the enzyme, thereby allowing the molecule to self-activate. Foltmann (1981) suggests that the gastric proteinases are held in an inactive form at neutral pH by electrostatic interactions between the basic amino acid residues of the propeptide and the negative dicarboxylic acid residues in the active site. Under acidic conditions, the carboxylic acid residues of the active site become protonated, and the molecule undergoes a conformational shift, rendering limited proteolytic activity. The enzymes are then irreversibly activated by proteolytic cleavage of the propeptide in a single- or multiple-step process.

The molecular mechanisms by which fungal aspartic proteinases are activated has remained relatively obscure. Two mechanisms have been proposed for the proteolytic removal of the propeptide: (a) The enzyme self-activates by intramolecular cleavage or (b) another proteinase acts on the enzyme to cleave the propeptide. Bovine chymosin, porcine pepsin, and proteinase A from *S. cerevisiae* are examples of self-activatable aspartic proteinases (Foltmann, 1981), while yeast proteinase B and renin are examples of proteinases that require other enzymes for activation (Mechler et al., 1987). Kodama (1988) has utilized active-site mutants of the aspartic proteinase from *M. miehei* in order to see if an inactive enzyme can be correctly processed in *A. nidulans*. It was reasoned that if the *Mucor* proteinase is a self-activating enzyme, one would expect that an inactive mutant gene would produce a gene product that could not cleave its own propeptide. However, if the *Mucor* proteinase zymogen was activated by another enzyme, one would expect an inactive mutant to be correctly processed. Utilizing site-directed mutagenesis, a strain of *A. nidulans* was isolated that produced an inactive form of *M. miehei* aspartic proteinase. Analysis of the aspartic proteinase produced by this strain revealed that the aminoterminal signal and propeptide sequences had been removed. Therefore, the processing of the zymogen in *A. nidulans* is most probably not an intramolecular event. When active *M. miehei* protease is expressed in *A. nidulans*, the amino-terminal residues are DGSVD. Amino-terminal sequencing of the inactive *Mucor* protein revealed an amino-terminal sequence of GSVD. Thus, the enzyme was processed, yet the amino-terminal end varied slightly from that of the expressed active enzyme. Whether this is due to improper processing by other enzymes or improper auto-activation is not yet known.

VII. CONCLUDING REMARKS AND FUTURE DIRECTIONS

Fungal aspartic proteinases have been used commercially for a number of industrial applications. In addition, these enzymes share several common features with some physiologically important mammalian aspartic proteinases. For these reasons, they have been the subject of vigorous biochemical and biophysical examination for several decades. Only recently have the techniques of modern molecular biology been employed to answer the remaining questions concerning structure-function relationships and post-translational processing of the aspartic proteinases. A popular approach has been to express the genes encoding the aspartic proteinases in a heterologous host organism for which genetic transformation procedures have been developed. One potential problem of studying proteolytic processing in heterologous fungal hosts is that many species produce aspartic proteinases of their own. This makes interpretation of data very difficult, since the activity of endogenous proteinases can interfere with experimental results. Thus, it would be very useful to have fungal host strains that were specifically lacking their native aspartic proteinases. Berka et al. (1990) have generated strains of *A. awamori* that contained deletions in the aspergillopepsin A structural gene. These strains were generated by transformation with linear DNA segments containing the aspergillopepsin A gene coding sequence interrupted by a selectable *argB* gene from *A. nidulans*. By a double cross-over event at the aspergillopepsin A gene locus, transformants were derived that produced no aspergillopepsin. These strains should prove to be useful hosts for studying the structure-function relationships of not only aspergillopepsin, but of other fungal aspartic proteinases as well. The powerful methods of site-directed mutagenesis will no doubt be employed to solve many unanswered questions about the molecular properties of these enzymes.

REFERENCES

Ammerer, G., Hunter, C.P., Rothman, J.H., Saari, G.C., Valls, L.A., and Stevens, T.H. (1986). *PEP4* gene of *Saccharomyces cerevisiae* encodes proteinase A, a vacuolar enzyme required for processing of vacuolar precursors. *Mol. Cell Biol.* *6*:2490-2499.
Andreeva, N., Zdanov, A., Gustchina, A., and Fedorov, A. (1985). X-ray diffraction analysis of porcine pepsin structure. In *Aspartic Proteinases and their Inhibitors*, Kostka, V., ed., Walter de Gruyter, Berlin, p. 137-150.
Arbige, M.V., Freund, P.R., Silver, S.C., and Zelko, J.T. (1986). Novel lipase for cheddar cheese flavor development. *Food Technol.* *40*:91-98.
Bakalinsky, A.T. and Boulton, R. (1985). The study of an immobilized acid protease for the treatment of wine proteins. *Am. J. Enol. Vitic.* *36*:23-29.
Ballance, D.J. and Turner, G. (1985). Development of a high-frequency transforming vector for *Aspergillus nidulans*. *Gene 36*:321-331.
Barkholt, V. (1987). Amino acid sequence of endothiapepsin: complete primary structure of the aspartic protease from *Endothia parasitica*. *Eur. J. Biochem.* *167*:327-338.
Bech, A.M. and Foltmann, B. (1981). Partial primary structure of *Mucor miehei* protease. *Neth. Milk Dairy J. 35*:275-280.
Berka, R.M., Ward, M., Wilson, L.J., Hayenga, K.J., Fong, K.K., Carlomagno, L.P., and Thompson, S.A. (1990). Molecular cloning and deletion of the gene encoding aspergillopepsin A from *Aspergillus awamori*. *Gene 86*:153-162.
Blundell, T., Jenkins, J., Pearl, L., Sewell, T., and Pedersen, V. (1985). The high resolution structure of endothiapepsin. In *Aspartic Proteinases and their Inhibitors,* Kostka, V., ed., Walter de Gruyter, Berlin, p. 151-161.
Blundell, T.L., Sewell, B.T., and McLachlan, A.D. (1979). Four-fold structural repeat in the acid proteases. *Biochim. Biophys. Acta 580*:24-31.
Bott, R., Subramanian, E., and Davies, D.R. (1982). Three dimensional structure of the complex of the *Rhizopus chinensis* carboxyl proteinase and pepstatin at 2.5 Angstrom resolution. *Biochemistry 21*:6956-6962.
Buxton, F.P., Gwynne, D.I., and Davies, R.W. (1985). Transformation of *Aspergillus niger* using the *argB* gene of *Aspergillus nidulans*. *Gene 37*:207-214.
Christensen, T., Woldike, H., Boel, E., Mortensen, S.B., Hjortshoj, K., Thim, L., and Hansen, M.T. (1987). High level expression of recombinant genes in *Aspergillus oryzae*. *Bio/Technology 6*:1419-1422.
Delaney, R., Wong, R.N.S., Meag, G., Wu, N., and Tang, J. (1987). Amino acid sequence of rhizopuspepsin isozyme pl 5. *J. Biol. Chem. 262*:1461-1467.

Dickinson, L., Harboe, M., Van Heeswijck, R., Stroman, P., and Jepsen, L.P. (1987). Expression of active *Mucor meihei* aspartic protease in *Mucor circinelloides*. *Carlsberg Res. Commun. 52*:243-252.

Foltmann, B. (1981). Gastric proteinases—structure, function, evolution and mechanism of action. *Essays Biochem. 17*:52-84.

Genencor Technical Bulletin R2 CL 84. *Rhozyme Proteases.* Genencor, Inc. South San Francisco, CA.

Graham, J.E.S., Sodek, J., and Hoffman, T. (1973). *Rhizopus* acid proteinases (*Rhizopus*-pepsins): properties and homology with other acid proteinases. *Can. J. Biochem. 51*:789-796.

Gray, G.L., Hayenga, K., Cullen, D., Wilson, L.J., and Norton, S. (1986). Primary structure of *Mucor miehei* aspartyl protease: evidence for a zymogen intermediate. *Gene 48*:41-53.

Hansen, J., Billich, S., Schulze, T., Sukrow, S., and Moelling, K. (1988). Partial purification and substrate analysis of bacterially expressed HIV protease by means of monoclonal antibody. *EMBO J. 7*:1785-1791.

Horiuchi, H., Yanai, K., Okazaki, T., Takagi, M., and Yano, K. (1988). Isolation and sequencing of a genomic clone encoding aspartic proteinase of *Rhizopus niveus J. Bacteriol. 170*:272-278.

Hsu, I., Delbaere, L., James, M., and Hofmann, T. (1977). Penicillopepsin from *Penicillium janthinellum* crystal structure at 2.8 Å and sequence homology with porcine pepsin. *Nature 266*:140-145.

Inagami, T., Misono, K., Chang, J., Takii, Y., and Dykes, C. (1985). Renin and general aspartyl proteases: differences and similarities in structure and function. In *Aspartic Proteinases and their Inhibitors*, Kostka, V., ed., Walter de Gruyter, Berlin, p. 319-337.

James, M., Sielecki, A., Salituro, F., Rich, D.H., and Hofmann, T. (1982). Conformational flexibility in the active sites of aspartyl proteinases revealed by a pepstatin fragment binding to penicillopepsin. *Proc. Natl. Acad. Sci. USA 79*: 6137-6141.

Kelly, J. and Hynes, M. (1985). Transformation of *Aspergillus niger* by the *amdS* gene. *EMBO J. 4*:475-479.

Kobayashi, H., Sekibata, S., Shibuya, H., Yoshida, S., and Kusakabe, I. (1989). Cloning and sequence analysis of cDNA for *Irpex lacteus* aspartic proteinase. *Agric. Biol. Chem. 53*:1927-1933.

Kodama, K.H. (1988). The expression of native and mutant *Mucor miehei* aspartic protease in *Aspergillus*. M.A. Thesis, San Francisco State University, San Francisco, CA.

Kyte, J. and Doolittle, R. (1982). A simple method for displaying the hydropathic character of a protein. *J. Mol. Biol. 157*:105-132.

Mechler, B., Muller, H., and Wolf, D.H. (1987). Maturation of vaculolar lysosomal enzymes in yeast proteinase yscA and proteinase yscB are catalysts of the processing and activation event of carboxypeptidase yscY. *EMBO J. 6*:2157-2163.

Miller, M., Jaskolski, M., Rao, J.K.M., Leis, J., and Wlodawer, A. (1989). Crystal structure of a retroviral protease proves relationship to aspartic protease family. *Nature 377*:576-579.

Navia, M.A., Fitzgerald, P.M.D., McKeever, B.M., Leu, C., Heimbach, J.C., Herber, W.K., Segal, I.S., Darke, P.L., and Springer, J.P. (1989). Three-dimensional structure of aspartyl protease from human immunodeficiency virus HIV-1. *Nature 337*:615-620.

Nevalainen, H., Penttilä, M., Harkki, A., Teeri, T., and Knowles, J. (1991). The molecular biology of *Trichoderma* and its application to the expression of both homologous and heterologous genes. This volume, p. 129.

Ostoslavskaya, V.I., Revina, L.P., Kotlova, E.K., Surova, I.A., Levin, E.D., Timokhina, E.A., and Stepanov, V.M. (1986). Primary structure of aspergillopepsin A—an aspartyl proteinase from *Aspergillus awamori*. *Bioorg. Khim. 12*:1030-1047.

Pearl, L.H., and Taylor, W.R. (1987). A structural model for the retroviral proteases. *Nature 329*:351-354.

Perlman, D., and Halvorson, H.O. (1983). A putative signal peptidase recognition site and sequence in eukaryotic and prokaryotic signal peptides. *J. Mol. Biol. 167*:391-409.

Rajogopalan, T.G., Stein, W.H., and Moore, S. (1966). Pepsin from pepsinogen. *J. Biol. Chem. 241*:4295-4297.

Sodek, J. and Hoffman, T. (1970). Large-scale preparation and some properties of penicillopepsin, the acid protease of *Penicillium janthinellum*. *Can. J. Biochem. 48*:415-431.

Tang, J. (1979). Evolution in the structure and function of carboxyl proteases. *Mol. Cell. Biochem. 26*:93-109.

Tang, J. and Wong, R.N.S. (1987). Evolution in the structure and function of aspartic proteases. *J. Cell. Biochem. 33*:53-63.

Tonouchi, N., Shoun, H., Uozumi, T., and Beppu, T. (1986). Cloning and sequencing of a gene for *Mucor* rennin, an aspartic protease from *Mucor pusillus*. *Nucleic Acids Res. 14*:7557-7568.

Umezawa, H., Aoyagi, T., Morishima, M., Matsuzaki, M., Hamada, M., and Takeuchi, T. (1970). Pepstatin, a new pepsin inhibitor produced by actinomycetes. *J. Antibiot. 32*:259-262.

Ward, M. (1991). Chymosin production in *Aspergillus*. This volume, p. 83.

Yamashita, T., Tonouchi, N., Uozumi, T., and Beppu, T. (1987). Secretion of *Mucor* rennin, a fungal aspartic protease of *Mucor pusillus*, by recombinant yeast cells. *Mol. Gen. Genet. 210*:462-467.

6

The Molecular Biology of *Trichoderma* and Its Application to the Expression of Both Homologous and Heterologous Genes

K. M. Helena Nevalainen
Alko Ltd., Helsinki, Finland

Merja E. Penttilä, Anu Harkki,* Tuula T. Teeri, and Jonathan Knowles†
The Technical Research Center of Finland, Espoo, Finland

I. INTRODUCTION

The mesophilic soft-rot fungi *Trichoderma*, in particular *Trichoderma reesei*, are without doubt the best-studied cellulolytic microorganisms. *Trichoderma* was "discovered" in Southeast Asia during the Second World War, where its ability to destroy cotton clothes and tents brought it to the attention of the U.S. Army. About 1950, Dr. Elwyn T. Reese and his co-workers of the U.S. Army Natick Laboratories identified *Trichoderma* strains producing a complete set of cellulase enzymes required for the breakdown of cellulose to glucose. In 1977, the *T. viride* strain Qm 6a, the ancestor of many mutant families developed throughout the world, was characterized as a distinct species of *T. longibrachiatum* group and was renamed *T. reesei* (Simmons, 1977).

Current affiliations:
*Technology Center, Cultor Ltd., Kantvik, Finland
†Glaxo Institute for Molecular Biology, Geneva, Switzerland

Although the work was initially directed towards inhibiting the cellulolytic activity of the fungus, the possibility of the production of cellulases
to bring about the enzymatic conversion of lignocellulose to glucose was
soon outlined by Mandels and Weber (1969). The oil shortage in the early
1970s stimulated a growing interest in alternative sources of fuels and chemicals, as well as in the study of cellulases. However, two economic bottle-
necks remained: the high costs of pretreatment of the lignocellulosic material for enzyme hydrolysis (especially delignification) and the high costs
of production of cellulases. One of the ways to overcome the problem was
to develop high-cellulase-producing *Trichoderma* mutant strains.

Some of these mutant strains produce up to 40 g/l of secreted protein,
most of which is cellobiohydrolase I. This represents well over half of the
total protein produced by the cell. The possibility of using molecular biology
to investigate this system and adapt it for the production of heterologous
proteins forms the background to the work reviewed here.

A. High-Producing Strains Developed by Traditional Mutagenesis and Screening Methods

Intensive strain development using the direct approach of mutation and
screening has successfully produced several high-yielding *T. reesei* mutant
strains (Table 1) in a number of laboratories. In the generation of hyper-
cellulolytic mutants, chemical mutagens, UV irradiation, X-rays, or mixed
treatments have been used. In the screening of individual colonies germinated from mutagen-treated spores for the production of improved amounts
of cellulases, phosphoric-acid-swollen Walseth-cellulose- (Walseth, 1952)
containing plates have proven to be useful. Major improvements have
been achieved in the total yield of cellulases, but only in two reports has

Table 1 Examples of *T. reesei* Hyperproducing Cellulase Mutant Series Isolated
Throughout the World

Mutant series	Reference
Natick strains, USA	(Mandels et al., 1971; Gallo et al., 1979; Andreotti et al., 1980)
VTT-D-series, Finland	(Nevalainen et al., 1980; Bailey and Neva-lainen, 1981)
Rutger's mutants, USA	(Montenecourt and Eveleigh, 1977, 1979; Montenecourt et al., 1979b)
L-series, Cetus Corporation, USA	(Shoemaker et al., 1981)
MHC-series, Czechoslovakia	(Farkaš et al., 1981; Labudová et al., 1981)
Kyowa strains, Japan	(Kawamori et al., 1985, 1986)
CL-mutants, France	(Durand et al., 1987)

the amount of an individual cellulase component, endoglucanase, been increased (Shoemaker et al., 1981; Sheir-Neiss and Montenecourt, 1984).

A number of different screening methodologies for the detection of regulatory cellulase mutants—especially catabolite repression-resistant and constitutive strains—have also been developed (Nevalainen and Palva, 1976; Montenecourt and Eveleigh, 1977; Montenecourt et al., 1979a; Montenecourt et al., 1979b; van Uden and Beja la Costa, 1980; Shoemaker et al., 1981; Nevalainen, 1985; Durand et al., 1987; Durand et al., 1984). Release from catabolite repression can result in the hyperproduction of enzyme(s). Indeed, many high-cellulase-producing *T. reesei* mutant strains have been reported to be less sensitive to glucose repression than the initial parent strains QM 6a or QM 9414 (e.g., Montenecourt et al., 1979b; Ruy et al., 1980; Shoemaker et al., 1981; Bailey and Oksanen, 1984).

However, none of the screening methods reported to date have sufficient precision to allow the selection of mutants of particular cellulase enzymes. This is in large part due to the complexity of the cellulolytic system produced by *Trichoderma*.

B. Molecular Biology Needed to Understand Cellulose Hydrolysis

As a substrate, cellulose is chemically very simple, consisting of a polymer of β-1,4-linked glucose units. However, it is physically a complex and variable structure. Moreover, in nature cellulose is associated with two other polymeric substances, lignin and hemicellulose. For this reason, large and complex batteries of different synergistically acting enzymes are needed for the natural recycling of lignocelluloses.

For hydrolysis of cellulose to glucose, three types of enzyme activities have been defined: endoglucanases (β-1,4-glucan glucanohydrolases, E.C. 3.2.1.4), exoglucanases (β-1,4-glucan cellobiohydrolases, E.C. 3.2.1.91), and β-glucosidase (β-D-glucoside glucohydrolase, E.C. 3.2.1.21). According to the generally accepted model for enzymatic cellulose degradation, endoglucanases (EG) cut internal β-1,4-glucosidic bonds in cellulose molecules, whereas cellobiohydrolases (CBH) cleave at the nonreducing end of the molecules and release the dissaccharide, cellobiose. β-Glucosidases hydrolyze cellobiose and other short oligosaccharides to glucose. However, refinements in the purification techniques and analysis techniques used to study the cellulases have led to new models for cellulose hydrolysis (Enari and Niku-Paavola, 1987). It is now clear that the mode of action of many of the *Trichoderma* cellulase cannot be classified as purely endo or exo type. The specificities of many of the cellulases may well be much broader.

Fortunately, in collaboration with Alwyn Jones at the University of Uppsala, Sweden, we have been able to crystallize both cellobiohydrolases, CBHI and CBHII (Bergfors et al., 1989). The three-dimensional structure of CBHII has already been solved at 2 Å resolution and the structure of CBHI is anticipated to be available soon. This information, coupled with studies on the mechanism of action, should enable us to understand more of the mechanism of enzymatic hydrolysis of cellulose in the not-too-distant future.

II. CELLULASE GENES

A. Induction of Cellulase Genes

Cellulolytic enzymes are generally induced synchronously (Eriksson and Goodell, 1974; Nevalainen and Palva, 1978; Durand et al., 1984), and their synthesis is subject to catabolite repression. However, the nature of the true inducer(s) triggering cellulase synthesis in nature or the mechanism by which the inducers regulate the transcription of the cellulase genes is not known at present. In laboratory experiments the substances used for the induction of cellulase synthesis are the disaccharides cellobiose and sophorose, which in nature may be produced by the activity of a β-glucosidase-type enzyme on the hydrolysis products of cellulose (Mandels and Reese, 1960; Mandels et al., 1962).

The expression of cellulase genes is regulated largely at the transcriptional level (Shoemaker et al., 1983; Teeri et al., 1983). Studies on the behavior of different cellulase negative mutants (Nevalainen and Palva, 1978; Durand et al., 1984) and the existence of different catabolite repression-insensitive mutants (Nevalainen and Palva, 1976; Montenecourt et al., 1979b; Shoemaker et al., 1981) suggest that production of the enzymes participating in the hydrolysis of cellulose is subject to multiple regulatory controls. Firstly, there seems to be an overall control system for the coordinate expression of cellulases, xylanase, mannanase, and β-glucosidase, as suggested by the existence of mutants not producing any of these enzymes. Secondly, β-glucosidase can be controlled separately from cellulase and hemicellulases. Evidence comes from the discovery of mutant strains lacking cellulases and hemicellulases but producing normal or elevated β-glucosidase activity. Thirdly, production of endoglucanases may be regulated separately from cellobiohydrolases: Mutant strains have been isolated in which the specific activity of endoglucanase has been increased. Durand et al. (1984) have also suggested an additional separate control system for xylanases, because they were able to isolate mutants that were xylanase positive, but cellulase and β-glucosidase negative, and produced evidence for common regulation of β-glucosidase and xylanase enzymes in *Trichoderma* (mutants that are cellulase negative, but xylanase and β-glucosidase positive). The nature of these postulated controls is not currently known but is being studied by applying recombinant DNA technology now available for *Trichoderma*.

B. Isolation and Characterization of Cellulase Genes

Four cellulase genes of *T. reesei* have been cloned and characterized in detail to date. These are two cellobiohydrolase genes, *cbh1* (Shoemaker et al., 1983; Teeri et al., 1983) and *cbh2* (Chen et al., 1987; Teeri et al., 1987), and two endoglucanase genes, *egl1* (Penttilä et al., 1986; van Arsdell et al., 1987) and *egl3* (Saloheimo et al., 1988).

The most successful method for the cloning of filamentous fungal cellulase genes has been to use differential cDNA hybridization of λ libraries of *T. reesei* (Shoemaker et al., 1983; Teeri et al., 1983; Penttilä et al., 1986). By applying two distinct cDNA probes synthesized using mRNA from mycelia grown on conditions either inductive or repressive for cellulase biosynthesis, genes highly expressed under induced conditions have been identified. Further characterization was carried out by hybrid selection of messenger RNA and/or sequence comparison to the available amino acid sequences of purified proteins.

C. Homology Between Different Cellulase Genes

The comparison of the predicted amino acid sequences derived from the four characterized cellulase genes (Figure 1.) revealed some extremely interesting areas of limited homology. These can be defined as follows. A region of 35 amino acids with 70% amino acid conservation is found in all the four *Trichoderma* cellulases. The cysteine and glycine residues are highly conserved. This homology block A is located at the C terminus in the cellulase enzymes endoglucanase I (EGI) and cellobiohydrolase I (CBHI) and at the N terminus in endoglucanase III (EGIII) and cellobiohydrolase II (CBHII) (Knowles et al., 1987a). A second homologous sequence B rich in Ser, Thr, Pro, and Arg is found next to block A in all four enzymes (Teeri et al., 1987). The block B sequence joins the A block to the main body of the enzyme and has been shown to be heavily 0 glycosylated in the CBHI (Fägerstam et al., 1984). The B block is apparently duplicated in CBHII, and there is evidence that most of the glycans are located in this region (Tomme et al., 1988).

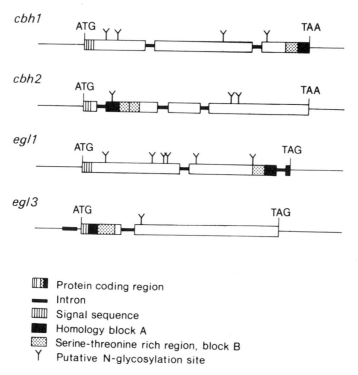

cbh1

cbh2

egl1

egl3

▥▤ Protein coding region
— Intron
▥ Signal sequence
■ Homology block A
▦ Serine-threonine rich region, block B
Y Putative N-glycosylation site

Figure 1   Schematic structure of four *T. reesei* cellulase genes.

Recent biochemical evidence supports the idea based on gene structure (Knowles, 1988) that cellulases are composed at least of two domains, a large isomorphic head and a more flexible terminal tail (Lehtovaara et al., 1986; Schmuck et al., 1986; Knowles et al., 1987b). The block B region might act as a flexible hinge linking the two domains together. Studies of truncated cellulases prepared by limited proteolysis of the *T. reesei* homologous terminal domains suggest a role for the terminal domains in substrate binding or solubilization (Knowles et al., 1987a; Knowles, 1988).

Enzymatic hydrolysis of cellulose almost certainly occurs by an acid-catalysis-type mechanism that is similar to that described for lysozymes. Limited homologies in the primary structures of cellulases to the active site of lysozymes, where the structure and functioning of the active site is already quite well understood, have been found (Paice et al., 1984; Teeri, 1987). However, in spite of the fact that the catalytic mechanisms are similar, it does not necessarily follow that the active-site structure should be identical.

Among the *T. reesei* cellulases, the deduced EGI protein with its signal peptide is 45% homologous to the CBHI amino acid sequence (Penttilä et al., 1986). CBHII and EGIII are both clearly different structurally, despite the fact that they both have the homologous conserved A block at their C terminus. Between the fungal cellulases, the *T. reesei* EGIII has been shown to be quite similar to the *Schizophyllum commune* EGI. The *S. commune* EGI as isolated, however, does not contain the AB block that is typical of the *T. reesei* cellulases characterized so far (Paice et al., 1984; Saloheimo et al., 1988). A *Phanerochaete chrysosporium* gene has been isolated by using a portion of a *Trichoderma cbh1* gene as a probe (Sims and Broda, 1987). Thus it would seem that at least some of the cellulase genes have evidently arisen by gene duplication accompanied by subsequent sequence divergence in the evolution of fungi.

D. Codon Usage in Cellulase Genes

The codon usage in *T. reesei* cellulase genes clearly differs from that of highly expressed yeast genes. For example, an A or T is preferred in the third position in yeast genes, but not in the open reading frame of the *T. reesei* gene *egl1* for endoglucanase (Penttilä et al., 1986). In the coding region of *egl1*, C and G are preferred in the third position of codons whenever possible (Pentillä et al., 1986). In the *T. reesei* cellobiohydrolase genes *cbh1* and *cbh2*, there is a strong bias against NTA codons where N is any nucleotide (Teeri et al., 1987). The putative splicing signal suggested for introns in filamentous fungi, $^T_A{}^{GCT}_C{}^{AAC}$ (Ballance, 1986), can be found in all introns of the *Trichoderma* cellulase genes (Figure 1).

Analysis of the 5' noncoding sequences of the *cbh1* and *cbh2* has revealed only one TATA-like sequence in the promoter region of both genes at nucleotide positions of -131 and -96 from the ATG, respectively (Shoemaker et al., 1983; Teeri et al., 1987). In addition, a possible CAAT box at position -187 from the ATG is found in *cbh2* (Teeri et al., 1987). As is typical for filamentous fungal genes, there are multiple start points for transcription of the *cbh2* gene, two of which are major. Other cellulase genes show a similar pattern of transcription initiation (Lehtovaara, unpublished data).

E. Expression of Cellulases in Heterologous Hosts

In the study of *Trichoderma* cellulases, the cDNAs of different cellulases have been expressed in yeast in order to study the biochemical properties of individual cellulase enzymes without the background given by the mixture of cellulases being present in the *Trichoderma* culture fluid.

The *T. reesei cbh1, cbh2, egl1*, and *egl3* cDNA copies were inserted into the yeast *PKG1* (Mellor et al., 1983) expression vector (Penttilä, 1987; Penttilä et al., 1987a; Penttilä et al., 1988). The expression of the cellulase genes in yeast was detected on plates containing different cellulosic substances, and all the four strains constructed secreted cellulases efficiently into the culture medium. Yeast seems to be able to secrete quite high amounts of *Trichoderma* cellulases (up to 100 mg/l of CBHII) (Penttilä, 1987; Penttilä et al., 1988). The proteins produced are overglycosylated, but nevertheless all are active towards their natural cellulosic substrates.

The expression of the major *T. reesei* cellulase gene *cbh1* has also been studied in *E. coli* (Teeri, 1987). The best expression levels were obtained when the mature *cbh1* was fused to a short N-terminal fragment of the phage λ Cro protein under the control of the lambda P_R promoter. However, the fusion protein formed intracellular aggregates that could be re-solubilized and renatured to restore the enzymatic activity on 4-methyl-β-umbelliferyl lactoside (Teeri, 1987).

Shoemaker and her colleagues at Genencor have expressed the *T. reesei cbh1* and *egl1* genes in *Aspergillus nidulans* (Barnett and Shoemaker, 1987). In addition, the *T. reesei cbh2* gene and modified forms of it have been expressed in *A. awamori* (Shoemaker et al., 1989). Interestingly, there seem to be differences in the posttranslational modification of the proteins in these filamentous fungi observed as heterogeneity in glycosylation. The effect of overglycosylation in yeast and *Aspergillus* or underglycosylation in *E. coli* on cellulase activity is still under investigation.

To further study the molecular biology of the cellulase genes and the mode of action of cellulase enzymes, it is clear that a transformation system for *Trichoderma* is required.

III. DEVELOPMENT OF A TRANSFORMATION SYSTEM IN *TRICHODERMA*

A. Auxotrophic Mutants

Well-defined auxotrophic mutant strains are not generally available in *T. reesei*. We have isolated and characterized auxotrophic mutants requiring uracil, tryptophan, or arginine using filtration enrichment (Fries, 1947) of UV-mutagenized conidia of *T. reesei* (Nevalainen, 1985; Penttilä et al., 1987b). Arginine-requiring mutant strains were enriched quite effectively, and mutants defective in the *arg B* gene (ornithine carbamoyl transferase) were sought by plating on MM glucose supplemented with ornithine, citrulline, or aginine. Two colonies requiring citrulline for growth (putative *arg B*) were chosen for transformation experiments. Tryptophan-requiring mutants were characterized on a series on MM plates supplied by anthranilic acid, indole, or tryptophan (Sanchez et al., 1987). Colonies requiring indole for growth were taken as putative *trpC* mutants.

B. Vector Systems and Transformation

In transformation of *T. reesei* auxotrophic mutants to prototrophy, vectors bearing *Aspergillus argB* (Penttilä et al., 1987b) and *trpC* genes (Nevalainen and Nyyssönen, unpublished data), as well as the *N. crassa pyr4* gene (Barnett and Shoemaker, 1987), have been used. No homologous marker system has been published for *Trichoderma* to date.

In addition to complementation of auxotrophy, dominant markers have been used in the transformation of filamentous fungi. Vectors carrying genes conferring resistance to aminoglycoside antibiotics, e.g., hygromycin B and G418, the metallo-glycopeptide antibiotic phleomycin, and the macrolide antibiotic oligomycin, have been constructed and applied (reviewed by Esser and Dohmen, 1987). Other dominant markers used in transformation include a gene giving an increase in resistance to the fungicide benomyl, a copper-resistance gene, and the *A. nidulans amdS* gene, which enables the organism to grow on acetamide as the sole nitrogen source (reviewed by Esser and Dohmen, 1987).

Transformatin of *T. reesei* has been carried out by using *amdS* (Penttilä et al., 1987b) and phleomycin (Durand et al., 1987). Insertion of *T. reesei* DNA fragments in front of a *Streptoallotheicus hindustanus* phleomycin resistance gene has given *T. reesei* transformant colonies resistant up to 1000 μg/ml of phleomycin (Durand et al., 1987). The method seems to be useful for cloning *T. reesei* fragments containing promoters. It has also been shown that the *E. coli lacZ* gene coupled with the *A. nidulans gpd* promoter (van Gorcom et al., 1986) is expressed in *T. reesei* and may thus be used in studies of promoter function in *T. reesei* (Penttilä et al., 1987b).

Transformation of *T. reesei* can be carried out by modifying the protocols developed for *Aspergillus* (Penttilä et al., 1987b). Protoplasts can be prepared either from cellophane-disc cultures or from shake-flask grown mycelia. In transformation, 2-5 μg (in cotransformation ~ 10 μg) of transforming DNA is used in 20 μl of buffer and mixed with 200 μl of protoplast suspension containing 5×10^6 to 5×10^7 protoplasts/ml. A solution of 25% PEG 6000 in 50 mM $CaCl_2$—10 mM Tris·HCl, pH 7.5, was used as the fusogenic agent.

The transformation frequencies obtained in *Trichoderma* are reasonably high: 150-400/μg with pSal43 (*argB*) and 50-400/μg with p3SR2 (*amdS*). The cotransformation efficiency with the pSal43 and p3SR2 in *Trichoderma* is high: 80-90% of the *argB* transformants were also Amd$^+$. Large Amd$^+$ Arg$^+$ colonies were obtained at the level of 50-100/μg and were accompanied by a number of small colonies typical for *amdS* transformation (Penttilä et al., 1987b).

C. Analysis and Stability of Transformants

Southern-blot analysis of a large number of ArgB$^+$ and AmdS$^+$ transformants indicates that integration of multiple tandem copies of transforming DNA occurs in different locations of the genome. The copy number of the transforming gene varied from 1 to about 20 (Penttilä et al., 1987b). Thus, a single copy of the *amdS* gene is sufficient to give a selectable phenotype in *T. reesei*.

Trichoderma AmdS$^+$ transformant colonies of various sizes were obtained. All the large colonies grew well when restreaked on acetamide-CsCl plates. Only about half of the smaller colonies grew well, so the other half were considered to be abortives. When sporulated on complete medium and retested for the AmdS$^+$ phenotype, some progeny of the original small colonies retained the AmdS$^+$ phenotype with a frequency varying from 5% to 94% of the individual spores tested and grew vigorously on acetamide. This indicates the occurrence of unstable transformants among small AmdS$^+$ colonies, in some of which, however, stable integration of transforming DNA may occur later and cause the restoration of growth.

The large *Trichoderma amdS* transformants showed a certain degree of mitotic instability. The *argB* transformants were, however, phenotypically stable for at least three generations. Three out of five *argB amdS* cotransformants tested for the stability of the ArgB$^+$ phenotype were shown to be 100% stable. In summary, transformation in *Trichoderma* shows many of the features described for other filamentous fungi.

IV. *TRICHODERMA* AS A PRODUCTION HOST FOR HOMOLOGOUS AND HETEROLOGOUS GENE PRODUCTS

A. The *T. reesei* Expression System

The best cellulase-producing *T. reesei* mutants can secrete up to 40 g of protein per liter under appropriate cultivation conditions. This represents over 50% of the total cell protein (Bailey and Nevalainen, 1981). Since over 60% of the secreted cellulase consists of CBHI enzyme, the product of a single *cbh1* gene, we have in hand a very strong promoter that directs the synthesis of about 30% of cellular protein. Moreover, it has been shown that the N-glycosylation pattern of the CBHI protein-bound glycans resembles closely that of the high mannose type found in certain animal cells (Salovuori et al., 1987). This suggests that *Trichoderma* could be an attractive host for the production of heterologous proteins.

The *cbh1* expression cassette (Figure 2) has been used for the production of calf chymosin in *T. reesei* (Harkki et al., 1987; Harkki et al., 1989) and in the construction of specific cellulase-producing *T. reesei* strains with tailored enzyme profiles (Harkki et al., 1990).

Proteases produced by the host may be harmful for heterologous gene products. The amount of proteases can be diminished by modifying

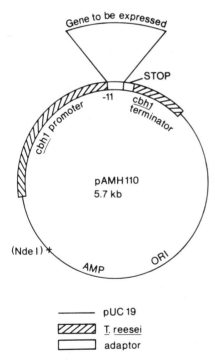

Figure 2 *Trichoderma reesei* expression vector pAMH110 containing the *cbh1* expression cassette.

the cultivation procedure and/or by isolating protease-negative mutant strains. Using UV mutagenesis we have isolated from *T. reesei* a mutant strain in which the production of acid protease is very much lower than the parent (Table 2). The mutation apparently has no effect on cellulase production.

The production of large amounts of endogenous CBHI might cause saturation of the secreted mechanism in strains producing heterologous proteins. To avoid this, cellobiohydrolase I negative (CBHI⁻) strains have been constructed by directed inactivation of the endogenous *cbhl* gene by disruption (Knowles et al., 1986). These CBHI⁻ strains clearly secreted less protein into the culture medium (Table 2) under the assay conditions and were shown not to produce CBHI.

B. Expression of *Aspergillus* Genes in *T. reesei*

Four *Aspergillus* promoter sequences, *argB, amdS, gpd* (Penttilä et al., 1987b), and *trpC* (Nevalainen and Nyyssönen, unpublished data) give rise to gene expression in *Trichoderma*. Since there are three introns in the *A. nidulans amdS* gene (Corric et al., 1987) and two introns in the N-terminal region of the *gpd* gene, it would also seem likely that these are processed by *T. reesei*.

Although the filamentous fungal genes seem to "cross-express," at least to some extent, in different heterologous filamentous fungal hosts, the level of expression of, e.g., the *A. nidulans gpd* promoter in *T. reesei*, seems to be low when compared to its activity in *Aspergillus*. In *Aspergillus*, transformant colonies carrying the *gpd-lacZ* fusions produce an intensive deep blue color when grown on plates containing the chromogenic substrate Xgal. In respective *Trichoderma* transformants, the color reaction is much less intense. Also, when the *T. reesei cbhl* gene was introduced in *A. nidulans*, the enzyme was produced in amounts only detectable in

Table 2 Selected Activities of Secreted Enzymes from *Trichoderma* in Shake-Flask Culture

Strain	Secreted protein (mg/ml)	Acid protease (U/ml)	Endoglucanase (U/ml)
Control	7.2	271	1510
Protease⁻	7.6	9.4	1510
Cellobiohydrolase I⁻	5.0	300	1470

Westerns (Penttilä, unpublished data). Thus, if efficient and regulated production of heterologous gene products in *T. reesei* is desired, the structural gene should be inserted into a *T. reesei* expression cassette.

C. Expression of Calf Chymosin in *T. reesei*

We are studying the expression of different mammalian proteins in *T. reesei*, the best studied case so far being calf chymosin. Four different expression plasmids have been constructed in which the chymosin cDNA was differently fused between the promoter and the terminator of the *T. reesei cbh1* gene (Harkki et al., 1987; Harkki et al., 1989). Fusions were made between the *cbh1* signal sequence and prochymosin cDNA, as well as using the *cbh1* promoter with the preprochymosin coding sequence. Plasmids carrying a signal fusion (amino acids 1-12 of CBHI, 12-16 of chymosin) and fusion in protein coding regions were also constructed, and the ability of *T. reesei* transformants carrying the different constructions to secrete calf chymosin into the growth medium was studied. The highest yield in initial fermentations, 40 mg/l, was obtained with a transformant carrying the plasmid pAMH106, which contains 20 amino acids of the mature CBHI protein fused to the prochymosin and the selection maker (*amdS*) in the same expression plasmid (Harkki et al., 1987; Harkki et al., 1989). The chymosin secreted by *T. reesei* was enzymatically active and the yield obtained is well in accordance or even more than that published for *A. nidulans* (Cullen et al., 1987) or *A. oryzae* (Boel et al., 1987). More recent experiments with protease-negative *T. reesei* hosts promise even greater yields.

D. Modulation of Homologous Gene Expression in *Trichoderma* by Promoter Exchange

Lignocellulose from different sources has widely varying properties and composition. For the optimum hydrolysis of a particular substrate, a specific mixture optimized for that substrate will clearly give the best results. Modulation of the production of *Trichoderma* enzymes has so far mainly been carried out using mutagenesis and screening for specific enzyme functions of the fungus. By conventional techniques, mutants producing elevated amounts of cellulases, strains lacking specific cellulase components, and protease-negative mutants have been induced and isolated, as discussed earlier. However, the use of random mutagenesis often leads to strains in which the production of all cellulases is improved, and also these strains are often unstable. Using cloned cellulase genes and their promoters, we can design and construct novel strains with completely different cellulase enzyme profiles.

1. Genetic Elimination of Particular Cellulase Activities

Strains lacking the major cellulase CBHI have been produced by specific inactivation of the gene (Knowles et al., 1986). Inactivation was carried out with a plasmid containing a 0.8-kb fragment from the 5′ terminal region of the *cbh1* gene, which also includes a frameshift mutation. The plasmid was introduced into *T. reesei* by cotransformation with the *amdS* gene. Transformants were first selected on the basis of the AmdS⁺ phenotype and purified. The cellulase phenotype was tested by Ouchterlony immunodiffusion from undiluted growth media against the CBHI-specific sheep antiserum. This led to the identification of a number of strains producing no detectable CBHI. The CBHI-negative nature of isolated strains was confirmed by analyzing the growth medium in FPLC in which no CBHI peak was seen. In the analysis of the specific messenger RNAs in Northern blots, small amounts of two CBHI⁻-specific truncated mRNAs were observed.

The lack of the major cellobiohydrolase did not interfere notably with the growth properties of the CBHI⁻ strain in fermenter cultivations. However, due to the lack of CBHI, the total amount of secreted protein was about half of that detected normally. This strain is propagated routinely on nonselective medium.

2. Elevating the Level of Specific Cellulases

We have also constructed *T. reesei* strains in which the production of endoglucanase is clearly improved. The *egl1* cDNA was inserted into the *cbh1* expression cassette of plasmid pAMH 110 (Figure 2) and cotransformed into *T. reesei* with the *amdS*-selective plasmid. Transformants were identified, purified, and 10 selected clones were tested for cellulase production in shake-flask cultures. The two best candidates were then grown in a laboratory fermenter and activity against hydroxyethylcellulose (HEC) was measured from the culture supernatants. One of these transformants produced significantly more endoglucanase activity than the control strain. A *cbh1* expression cassette containing *cbh2* cDNA has also been introduced in *T. reesei* to improve the production of this particular cellulase and, moreover, strains lacking the CBHII enzyme protein have been constructed recently (own unpublished data). These results show that genetic engineering can be used to modulate gene expression in *T. reesei* and to produce new strains of considerable commerical interest that produce novel cellulase preparations with radically modified enzyme composition.

V. CONCLUSIONS

A. Novel Cellulase Mixtures

We have here reviewed the different aspects of the molecular biology of *Trichoderma* that now make it possible to carry out elaborate genetic manipulation of this organism. Thus the lack of a mating system in *Fungi imperfecti* such as *Trichoderma* is no longer a serious handicap. The study of the molecular biology of *Trichoderma* has provided interesting fundamental information about the organism and has now made possible the construction of *Trichoderma* strains of significant commercial importance.

We have shown that specific genes can be inactivated or their expression can be increased, which results in strains capable of producing specifically tailored enzyme cocktails for different uses. Also, the production of single cellulase components is possible in genetically engineered *T. reesei* strains.

T. reesei cellulases are presently used in grain alcohol fermentation, starch processing, animal feed applications, malting and brewing, and in the production of fruit and vegetable juices. New promising areas being intensively studied are the use of cellulases in waste-water treatment and in different pulp and paper production processes.

In the future other heterologous enzyme genes from various sources will be expressed in *Trichoderma* to further improve the industrial properties of *Trichoderma* enzyme mixtures. Specific modifications of lignocellulose will become as important as its hydrolysis in the future. Protein engineering will be used to improve the thermoresistance and to reduce product inhibition of the *Trichoderma* cellulases, and this may extend further the possible applications of the *Trichoderma* enzymes.

B. *Trichoderma* as a System for the Production of Heterologous Proteins

Currently, animal cell culture is often the only way to produce pharmaceutically and commercially important proteins. The increasing problems associated with the purification of proteins from blood serum caused by the rapid spread of pathogenic viruses make it even more important to develop new production systems.

Filamentous fungi and *Trichoderma*, in particular, would seem to offer a very reasonable alternative for the production of secreted mammalian proteins. The production of 40 mg/l of calf chymosin by one of the first *T. reesei* transformants studied (Harkki et al., 1987; Harkki et al., 1989) support this view. The use of protease-negative hosts and more refined DNA constructions has, in our hands, given even greater yields of chymosin from *Trichoderma*. It would seem that the chymosin produced in *Trichoderma* is not glycosylated, as judged from the mobility of the product in SDS-PAGE gels.

In summary, the application of genetic engineering to the cellulase-producing fungus *T. reesei* has increased our understanding of the molecular biology of this important organism. In addition, the application of this knowledge has made possible the construction of a number of strains with considerable commercial potential. In the future, we believe that *Trichoderma* will prove increasingly useful for the production of both bulk enzymes and of a variety of pharmaceutical products.

REFERENCES

Andreotti, R., Medeiros, J., Roche, C., and Mandels, M. (1980). Effects of strain and substrate on production of cellulases by *Trichoderma reesei* mutants. In *Proceedings of Second International Symposium on Bioconversion and Biochemical Engineering*, Ghose, T.K., ed., ITT, Delhi, p. 353-388.

van Arsdell, J.N., Kwok, S., Schweickart, V.L., Ladner, M.B., Gelfand, D.H., and Innis, M.A. (1987). Cloning, characterization, and expression in *Saccharomyces cerevisiae* of endoglucanase I from *Trichoderma reesei*. *Bio/Technology* 5:60-64.

Bailey, M.J. and Nevalainen, K.M.H. (1981). Induction, isolation and testing of stable *Trichoderma reesei* mutants with improved production of solubilizing cellulase. *Enzyme Microb. Technol. 3*:153-157.

Bailey, M.J. and Oksanen, J. (1984). Cellulase production by mutant strains of *Trichoderma reesei* on noncellulosic media. In *Proceedings of the 3rd European Congress on Biotechnology*, Vol. II, München, p. 157-162.

Ballance, J.B. (1986). Sequences important for gene expression in filamentous fungi. *Yeast 2*:229-236.

Barnett, C. and Shoemaker, S.P. (1987). Expression of endoglucanase genes of *Trichoderma reesei* in *Aspergillus nidulans*. In *FEMS Symposium on the Biochemistry and Genetics of Cellulose Degradation*, Paris, Poster 2-18.

Bergfors, T., Rouvinen, J., Lehtovaara, P., Caldentey, X., Tomme, P., Claeyssens, M., Pettersson, G., Teeri, T., Knowles, J., and Jones, T.A. (1989). Crystallization of the core protein of cellobiohydrolase II from *T. reesei*. *J. Mol. Biol. 209*:167-169.

Boel, E., Christensen, T., and Wöldike, H.F. (1987). Process for the production of protein products in *Aspergillus oryzae* and a promoter for use in *Aspergillus*. *European Patent Application* #87103806.3.

Chen, C.M., Gritzali, M., and Stafford, D.W. (1987). Nucleotide sequence and deduced primary structure of cellobiohydrolase II of *Trichoderma reesei*. *Bio/Technology 5*:274-278.

Corric, C.M., Twomey, A.P., and Hynes, M.J. (1987). The nucleotide sequence of the *amdS* gene of *Aspergillus nidulans* and the molecular characterization of 5' mutations. *Gene 53*:63-71.

Cullen, D., Gray, G.L., Wilson, L.J., Hayenga, K.J., Lamsa, M.H., Rey, M.W., Norton, S., and Berka, R.M. (1987). Controlled expression and secretion of bovine chymosin in *Aspergillus nidulans*. *Bio/Technology 5*:369-376.

Durand, H., Baron, M., Calmels, T., and Tiraby, G. (1987). Classical and molecular genetics applied to *Trichoderma reesei* for the selection of improved cellulolytic industrial strains. In *Biochemistry and Genetics of Cellulose Degradation*, Aubert, J.-P., Beguin, P., and Millet, J., eds., Academic Press, p. 135-151.

Durand, H., Clanet, M., and Tiraby, G. (1984). A genetic approach of the improvement of cellulase production by *Trichoderma reesei*. In *Proceedings of Bioenergy World Conference*, Gothenburg, Sweden, Egneus, H. and Ellegard, A., eds., Elsevier Applied Sciences Publishers, London, Vol. 3, pp. 246-253.

Enari, T.-M. and Niku-Paavola, M.-L. (1987). Enzymatic hydrolysis of cellulose: is the current theory of hydrolysis valid? *CRC Crit. Rev. Biotechnol. 5*:67-87.

Eriksson, K.E. and Goodell, E.W. (1974). Pleiotropic mutants of the wood-rotting fungus *Polyporus adustus. Can. J. Microbiol. 20*:371-378.

Esser, K. and Dohmen, G. (1987). Drug resistance genes and their use in molecular cloning. *Process Biochem. 22*:144-147.

Fägerstam, L., Pettersson, L.G., and Engström, J.Å. (1984). The primary structure of a 1,4-β-glucan cellobiohydrolase from the fungus *Trichoderma reesei* QM9414. *FEBS Lett. 167*:309-315.

Farkaš, V., Labudová, I., Bauer, Š, and Ferenczy, L. (1981). Preparation of mutants of *Trichoderma viride* with increased production of cellulase. *Folia Microbiol. 26*:129-132.

Fries, N. (1947). Experiments with different methods of isolating physiological mutations of filamentous fungi. *Nature 159*:199.

Gallo, B.J., Andreotti, R., Roche, C., Ruy, D., and Mandels, M. (1979). Cellulase production by a new mutant strain of *Trichoderma reesei* MCG 77. *Biotechnol. Bioengineer. Symp. 8*:89-102.

van Gorcom, R.F.M., Punt, P.J., Pouwels, P.H., and van den Hondel, C.A.M.J.J. (1986). A system for the analysis of expression signals in *Aspergillus*. *Gene 48*: 211-217.

Harkki, A., Bailey, M., Penttilä, M., and Knowles, J. (1987). Active calf chymosin efficiently secreted from *Trichoderma reesei*. In *Abstracts of the 19th Lunteren Lectures on Molecular Genetics*, Lunteren, p. 42.

Harkki, A., Mäntylä, A., Penttilä, M., Muttilainen, S., Bühler, R., Knowles, J., and Nevalainen, H. (1990). Genetic engineering of *Trichoderma* to produce strains with novel cellulase profiles. *Enzyme Microb. Technol.*, in press.

Harkki, A., Uusitalo, J., Bailey, M., Penttilä, M., and Knowles, J.K.C. (1989). Efficient secretion of active calf chymosin from the filamentous fungus *Trichoderma reesei. Bio/Technology 7*:596-603.

Kawamori, M., Morikawa, Y., Shinsa, Y., Takayama, K., and Takasawa, S. (1985). Preparation of mutants resistant to catabolite repression of *Trichoderma reesei. Argric. Biol. Chem. 49*:2875-2879.

Kawamori, M., Ado, Y., and Takasawa, S. (1986). Preparation and application of *Trichoderma reesei* mutants with enhanced β-glucosidase. *Agric. Biol. Chem. 50*:2477-2482.

Knowles, J. (1988). The use of gene technology to investigate fungal cellololytic enzymes. In *Biochemistry and Genetics of Cellulose Degradation*, Aubert, J.-P., Beguin, P., and Millet, J., eds., Academic Press, New York, p. 153-169.

Knowles, J., Harkki, A., Penttilä, M., Nevalainen, H., and Hansen, M.T. (1986). Transformation of *Trichoderma. European Patent Application #87303834.3.*

Knowles, J., Lehtovaara, P., and Teeri, T. (1987a). Cellulase families and their genes. *Tibtech. 5*:255-261.

Knowles, J., Lehtovaara, P., Teeri, T., Penttilä, M., Salovuori, I., and André, L. (1987b). The application of recombinant-DNA technology to cellulases and lignocellulosic wastes. *Phil. Trans. R. Soc. Lond. A321*:449-454.

Labudová, I., Farkaš, V., Bauer, Š., Kolarová, N., and Brányik, A. (1981). Characterization of cellulolytic enzyme complexes obtained from mutants of *Trichoderma reesei* with enhanced cellulase production. *Eur. J. Appl. Microbiol. Biotechnol. 21*:16-21.

Lehtovaara, P., Knowles, J., André, L., Penttilä, M., Teeri, T., Salvovuori, I., Niku-Paavola, M.-L., and Enari, T.-M. (1986). The application of recombinant DNA for the production of pure enzymes for the pulp and paper industry. In *The Third International Conference on Biotechnology in the Pulp and Paper Industry*, Stockholm, p. 90-92.

Mandels, M. and Reese, E.T. (1960). Induction of cellulase by cellobiose. *J. Bacteriol. 79*:816-826.

Mandels, M., Parrish, F.W., and Reese, E. (1962). Sophorose as an inducer of cellulase in *Trichoderma viride*. *J. Bacteriol. 83*:400-408.

Mandels, M. and Weber, J. (1969). The production of cellulases. *Adv. Chem. Ser. 95*:391-403.

Mandels, M., Weber, J., and Parizek, R. (1971). Enhanced cellulase production by a mutant of *Trichoderma viride*. *Appl. Microbiol. 21*:152-154.

Mellor, J., Dobson, M.J., Roberts, N.A., Tuite, M.F., Emtage, J.S., White, S., Lowe, P.A., Patel, T., Kingsman, A.J., and Kingsman, S.M. (1983). Efficient synthesis of enzymatically active calf chymosin in *Saccharomyces cerevisiae*. *Gene 24*:1-14.

Montenecourt, B.S. and Eveleigh, D.E. (1977). Preparation of mutants of *Trichoderma reesei* with enhanced cellulase production. *Appl. Environ. Microbiol. 34*:777-782.

Montenecourt, B.S. and Eveleigh, D.E. (1979). Selective isolation of high yielding cellulase mutants of *T. reesei*. *Adv. Chem. Ser. 181*:289-301.

Montenecourt, B.S., Schamhart, D.H.J., Cuskey, S.M., and Eveleigh, D.E. (1979a). Improvements in cellulase production by *Trichoderma* through mutation and selected breeding. In *Proceedings of 3rd Annual Biomass Energy Systems Conference*, p. 85-89.

Montenecourt, B.S., Schamhart, D.H.J., and Eveleigh, D.E. (1979b). Mechanisms controlling the synthesis of the *Trichoderma reesei* cellulase system. In *Microbial Polysaccharides and Polysaccharases*, Berkeley, R.C.W., Gooday, G., and Ellwood, D.C., eds., Academic Press, New York, p. 327-337.

Nevalainen, H. (1985). *Genetic Improvement of Enzyme Production in Industrially Important Fungal Strains*. Technical Research Centre of Finland, Espoo Publication 26.

Nevalainen, H. and Palva, T. (1976). The regulation of cellulase, cellobiase, xylanase and mannanase production in *Trichoderma viride*. In *Meeting Abstracts of the 8th North West European Microbial Group*, Helsinki, p. 114.

Nevalainen, K.M.H. and Palva, E.T. (1978). Production of extracellular enzymes in mutants isolated from *Trichoderma viride* unable to hydrolyze cellulose. *Appl. Environ. Microbiol. 35*:11-16.

Nevalainen, K.M.H., Palva, E.T., and Bailey, M.J. (1980). A high cellulase-producing mutant strain of *Trichoderma reesei*. *Enzyme Microb. Technol. 3*:59-60.

Paice, M.G., Desrochers, D.R., Jurasek, C.R., Rollin, C.F., De Miguel, E., and Yaguchi, M. (1984). Two forms of endoglucanase from the basidiomycete *Schizophyllum commune* and their relationship to other β-1,4-glycoside hydrolases. *Bio/Technology 2*:535-539.

Penttilä, M. (1987). *Construction and Characterization of Cellulolytic Yeasts.* Technical Research Centre of Finland, Espoo Publication 39.

Penttilä, M.E., André, L., Lehtovaara, P., Bailey, M., Teeri, T., and Knowles, J.K.C. (1988). Efficient secretion of two fungal cellobiohydrolases in *Saccharomyces cerevisiae. Gene 63*:103-112.

Penttilä, M., Lehtovaara, P., Nevalainen, H., Bhikhabhai, R., and Knowles, J. (1986). Homology between cellulase genes of *Trichoderma reesei*: complete nucleotide sequence of the endoglucanase I gene. *Gene 45*:253-263.

Penttilä, M.E., André, L., Saloheimo, M., Lehtovaara, P., and Knowles, J.K.C. (1987a). Expression of two *Trichoderma reesei* endoglucanases in the yeast *Saccharomyces cerevisiae. Yeast 3*:175-185.

Penttilä, M., Nevalainen, H., Rättö, M., Salminen, E., and Knowles, J. (1987b). A versatile transformation system for the cellulolytic filamentous fungus *Trichoderma reesei. Gene 61*:155-164.

Ruy, D.D.Y., Andreotti, R., Medeiros, J., and Mandels, M. (1980). Comparative quantitative physiology of high cellulase producing strains of *Trichoderma reesei. Enzyme Engineer. 5*:33-40.

Saloheimo, M., Lehtovaara, P., Penttilä, M., Teeri, T.T., Ståhlberg, J., Johansson, G., Pettersson, G., Claeyssens, M., Tomme, P., and Knowles, J. (1988). A new endoglucanase from *Trichoderma reesei*: the characterization of both gene and enzyme. *Gene 63*:103-112.

Salovuori, I., Makarow, M., Rauvala, H., Knowles, J., and Kääriäinen, L. (1987). Low molecular weight high-mannose type glycans in a secreted protein of the filamentous fungus *Trichoderma reesei. Bio/Technology 5*:152-156.

Sanchez, F., Lozano, M., Rubio, V., and Peñalva, M.A. (1987). Transformation in *Penicillium chrysogenum. Gene 51*:97-102.

Schmuck, M., Pilz, I., Hayn, M., and Esterbauer, H. (1986). Investigation of cellobiohydrolase from *Trichoderma reesei* by small x-ray scattering. *Biotechnol. Lett. 8*:397-402.

Sheir-Neiss, G. and Montenecourt, B.S. (1984). Characterization of the secreted cellulases of *Trichoderma reesei* wild type and mutants during controlled fermentations. *Appl. Microbiol. Biotechnol. 20*:46-53.

Shoemaker, S., Barnett, L., Sumner, L., Jacobs, R., Berka, R., and Pover, S. (1989). Cellulose depolymerases of *Trichoderma*: nomenclature and properties of some recombinant forms. In Proceedings of the Tricel 89 International Symposium, Vienna, in press.

Shoemaker, S.P., Raymond, J.C., and Bruner, R. (1981). Diversity amongst improved *Trichoderma* strains. In *Trends in the Biology of Fermentations for Fuels and Chemicals,* Hollaender, A.E., ed., Plenum Press, New York, p. 89-109.

Shoemaker, S., Schweickart, V., Ladner, M., Gelfand, D., Kwok, S., Myambo, K., and Innis, M. (1983). Molecular cloning of exocellobiohydrolase I derived from *Trichoderma reesei* strain L27. *Bio/Technology 1*:691-696.

Simmons, E.G. (1977). Classification of some cellulase producing *Trichoderma* species. In *Abstracts of Second International Mycological Congress*, Tampa, FL, Aigelow, H.E., and Simmons, E.G., eds., p. 168.

Sims, P. and Broda, P. (1987). The identification, molecular cloning and characterization of a gene from *Phanerochaete chrysosporium* that shows strong homology to the exo-cellobiohydrolase I gene from *Trichoderma reesei*. In *FEMS Symposium Biochemistry and Genetics of Cellulose Degradation*, Paris, Poster 24.

Teeri, T. (1987). *The Cellulolytic Enzyme System of Trichoderma reesei. Molecular Cloning, Characterization and Expression of the Cellobiohydrolase Genes.* Technical Research Centre of Finland, Espoo Publication 38.

Teeri, T., Salovuori, I., and Knowles, J. (1983). The molecular cloning of the major cellulase gene from *Trichoderma reesei*. *Bio/Technology 1*:696-699.

Teeri, T., Lehtovaara, P., Kauppinen, S., Salovuori, I., and Knowles, J. (1987). Homologous domains in *Trichoderma reesei* cellulolytic enzymes: gene sequence and expression of cellobiohydrolase II. *Gene 51*:43-52.

Tomme, P., van Tilbeurgh, H., Pettersson, G., van Damme, J., Vandekerchove, J., Knowles, J., Teeri, T., and Claeyssens, M. (1988). Studies of the cellulolytic system of *Trichoderma reesei* QM9414. Analysis of domain function in two cellobiohydrolases by limited proteolysis. *Eur. J. Biochem. 170*:575-581.

van Uden, N. and Beja la Costa, M. (1980). Use of 2-deoxyglucose in the selective isolation of mutants of *Trichoderma reesei* with enhanced β-glucosidase production. *Biotechnol. Bioeng. 22*:2429-2432.

Walseth, C.S. (1952). Occurrence of cellulases in enzyme preparations from microorganisms. *T.A.P.P.I. 35*:228-233.

7

Molecular Genetics of Penicillin and Cephalosporin Antibiotic Biosynthesis

Juan F. Martin
University of León, León, Spain

Thomas D. Ingolia and Stephen W. Queener
Lilly Research Laboratories, Indianapolis, Indiana

I. BIOCHEMICAL GENETICS OF PENICILLIN AND CEPHALOSPORIN BIOSYNTHESIS

Application of molecular genetics to antibiotic biosynthesis and production in microorganisms has increased rapidly in the last few years (Martin and Gil, 1982; Hopwood and Chater, 1984; Ingolia and Queener, 1989; Miller and Ingolia, 1989) but is still in the early stages of development. The medical and industrial importance of these compounds insures further progress. The biosynthesis of penicillins and cephalosporins is relatively well understood; steps in the biosynthetic pathways of penicillin G (and V) and of several cephalosporins have been characterized at the enzyme level (Queener and Neuss, 1982; Demain, 1983; Martin and Aharonowitz, 1983; Abraham, 1986; Martin and Liras, 1985; Baldwin and Abraham, 1988.)

A. Common Steps in the Biosynthesis of Penicillins and Cephalosporins

The bicyclic *penam* ring system — beta-lactam ring fused to a thiazolidine ring — is present in penicillins. The bicyclic *cephem* ring system — beta-lactam fused to a dihydrothiazine ring — is present in cephalosporins and cephamycins (7-alpha-methoxycephalosporins). The penam and cephem ring systems are derived from L-cysteine and L-valine. A third amino acid, L-alpha-aminoadipic acid, is required to biosynthesize penicillins and cephalosporins; it becomes the alpha-aminodipyl portion of L- or D-alpha-aminoadipamido moieties attached at C6 of the penam or C7 of the cepham ring systems. These natural side chains are attached to the beta face of the ring systems.

1. Formation of ACV

L-alpha-aminoadipic acid, L-cysteine, and L-valine are required to form the tripeptide delta-(L-alpha-aminoadipyl)-L-cysteinyl-D-valine (ACV) (Fawcett et al., 1976), an intermediate in the biosynthesis of all pencillins and cephalosporins. The enzyme catalyzing this reaction has been referred to as *ACV synthetase*. ATP is required for the reaction and a "ligase" mechanism is probably involved (Banko et al., 1987). The enzyme from *Aspergillus nidulans* has been studied; one fraction not only formed ACV from the constituent amino acids, but also activated the constituent amino acids as their adenylates (van Liempt et al., 1989). Thus, ACVS in *A. nidulans* appears to be a single enzyme. The enzyme originally reported to have a molecular weight of ca. 220 kDa, upon further analysis, appears larger, ca. 285 kDa (van Liempt, personal communication).

2. Cyclization of ACV

ACV is cyclized to form isopenicillin N (IPN), an intermediate that can be visualized as an L-alpha-aminoadipyl side chain attached via amide linkage to a 6-aminopenicillanic acid (6-APA) nucleus (Figure 1). Cell-free extracts of *A. chrysogenum* (Fawcett et al., 1976; Pang et al., 1984), *Streptomyces clavuligerus* (Jensen et al., 1982), and *P. chrysogenum* (Ramos et al., 1985) catalyze the reaction. The enzyme catalyzing the cyclization has been designated by the trivial designations *cyclase, isopenicillin N synthetase,* or *isopenicillin N synthase* (the latter two abbreviated as IPNS). *Synthetase* and *synthase* imply *ligase* and *lyase* mechanisms, respectively. However, ATP is not required for the reaction and no double bonds are formed; there is no evidence that either mechanism occurs in the cyclization. The cyclization of ACV to IPN involves an "oxidoreductase" mechanism: Four hydrogen atoms are removed from ACV and one molecule of oxygen is consumed (Baldwin et al., 1982). ACV: oxygen oxidoreductase (cyclizing) may be appropriate as a systematic name, but is cumbersome. In this review we use the name *ACV cyclase* or simply *cyclase*. This unambiguously designates the enzyme without implying an unlikely reaction mechanism.

ACV cyclase from *P. chrysogenum* (Ramos et al., 1985; Martin et al., 1987), *A. chrysogenum* (syn. *Cephalosporium acremonium, Acremonium strictum*) (Hollander et al., 1984), *S. clavuligerus* (Jensen et al., 1982), *N. lactamdurans* (Castro et al., 1988), and *S. lipmanii* (Weigel et al., 1988) have each been purified (M_r 39,000, 41,000, 31,000, 26,500, 38,000, respectively). In all five microorganisms, cyclization requires dithiothrietol and is stimulated by ferrous ions and ascorbate (see review by Martin and Liras, 1985).

In *P. chrysogenum* and *A. nidulans*, IPN is required for the biosynthesis of precursable penicillins, e.g., penicillin G and V. In *A. chrysogenum, S. clavuligerus, N. lactamdurans,* and *S. lipmanii*, IPN is required for the biosynthesis of cephalosporins, e.g., cephalosporin C and cephamycin C.

B. Biosynthesis of Precursable Penicillins: Transacylation of IPN

P. chrysogenum and *A. nidulans* produce penicillins with acyl side chains formed from a variety of intracellular or exogenously supplied carboxylic acids. When exogenous phenylacetic acid (PA) or phenoxyacetic acid (PoA) is provided, penicillins with phenylacetyl and phenoxyacetyl side chains, e.g., penicillin G and V (Figure 1), are produced, respectively. Penicillins formed in this manner are said to be "precursable." Acids acceptable for "precursing" have nonpolar side chains and a free methylene moiety adjacent to the carboxyl. The acyl moiety of such acids are believed to be incorporated into penicillins via activated derivatives, particularly coenzyme A derivatives. Extracts of *P. chrysogenum* catalyze the formation of phenylacetylCoA (PACoA) and phenoxyacetylCoA (PoACoA) from PA or PoA, respectively, in the presence of, added coenzyme A and ATP (Brunner et al., 1968). A number of monosubstituted acetic acids were equally active in the reaction. The concept of a monosubstituted acetic acid ligase in *P. chrysogenum* governed by single gene (*pen*C) has been suggested (Ingolia and Queener, 1989). Purification of such an enzyme and cloning of its gene have not yet been demonstrated.

The biosynthesis of penicillin G and V (Figure 1) occurs by replacement of the L-alpha-aminoadipyl side chain of IPN with the phenylacetyl moiety of phenylacetylCoA (PACoA) or the phenoxyacetyl moiety of phenoxyacetylCoA (PoACoA), respectively. The conversions are generally believed to take place via a two-step reaction in which 6-APA is formed and then acylated without release from the enzyme surface.

The enzyme involved, acylCoA:IPN acyltransferase, appears to contain at least three types of binding sites: (a) an activated acyl binding site, which

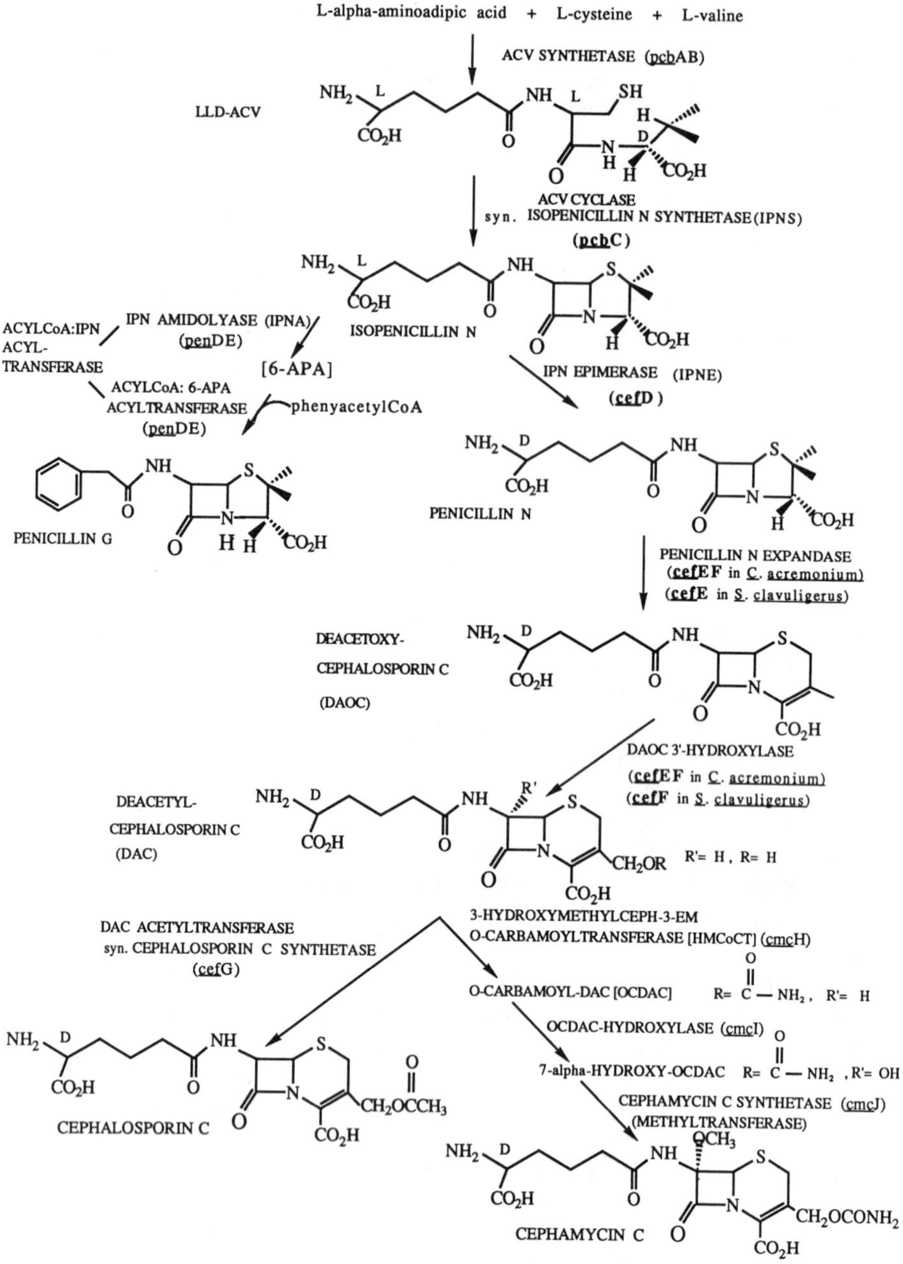

L-alpha-aminoadipic acid + L-cysteine + L-valine

ACV SYNTHETASE (pcbAB)

LLD-ACV

ACV CYCLASE
syn. ISOPENICILLIN N SYNTHETASE (IPNS)
(pcbC)

ISOPENICILLIN N

ACYLCoA:IPN ACYL-TRANSFERASE

IPN AMIDOLYASE (IPNA) (penDE)

ACYLCoA: 6-APA ACYLTRANSFERASE (penDE)

[6-APA]

phenyacetylCoA

IPN EPIMERASE (IPNE)
(cefD)

PENICILLIN G

PENICILLIN N

PENICILLIN N EXPANDASE
(cefEF in C. acremonium)
(cefE in S. clavuligerus)

DEACETOXY-CEPHALOSPORIN C
(DAOC)

DAOC 3'-HYDROXYLASE
(cefEF in C. acremonium)
(cefF in S. clavuligerus)

DEACETYL-CEPHALOSPORIN C
(DAC)

R'= H , R= H

DAC ACETYLTRANSFERASE
syn. CEPHALOSPORIN C SYNTHETASE
(cefG)

3-HYDROXYMETHYLCEPH-3-EM
O-CARBAMOYLTRANSFERASE [HMCoCT] (cmcH)

O-CARBAMOYL-DAC [OCDAC]   R= $C-NH_2$, R'= H

OCDAC-HYDROXYLASE (cmcI)

7-alpha-HYDROXY-OCDAC   R= $C-NH_2$, R'= OH

CEPHALOSPORIN C

CEPHAMYCIN C SYNTHETASE (cmcJ)
(METHYLTRANSFERASE)

CEPHAMYCIN C

interacts with various activated forms of phenylacetyl and phenoxyacetyl moieties, i.e., PACoA, penicillin G, PoACoA, and penicillin V; (b) a binding site for IPN; and (c) one or more binding sites for 6-APA. The organization of these binding sites is not understood, but several observations pertinent to understanding the gross structure of the enzyme have been made.

N-terminal and internal amino acid sequences in the 10- and 30-kDa proteins (Barredo et al., 1989b; Veenstra et al., 1989; Whiteman et al., 1990) were shown to be encoded by a single open reading frame in libraries of *P. chrysogenum* DNA (Veenstra et al., 1989; Tobin, Fleming, Skatrud, Miller, personal communication). That orf, taken from a cDNA library and expressed in *E. coli*, produced a 40-kDa polypeptide with acyl-CoA:IPN acyltransferase activity (Tobin, Skatrud, and Miller, personal communication). Separation of the 10- and 30-kDa proteins—purified from *P. chrysogenum*—was associated with almost complete loss of acyl-CoA:6-APA acyltransferase activity (see below), which could be stimulated 15-fold by mixing the two proteins (Whiteman et al., 1990).

Thus, the primary functional form of acylCoA:IPN acyltransferase in *P. chrysogenum* appears to be a heterodimer composed of a 10-kDa and a 30-kDa protein. A single open reading frame appears to encode a 40-kDa protein, which is processed in *P. chrysogenum* to 10- and 30-kDa subunits. In *E. coli*, domains corresponding to the 10- and 30-kDa subunits probably exist and function within the 40-kDa polypeptide that has acylCoA:IPN acyltransferase activity.

There are several possibilities for distribution of binding sites on the two subunits in *P. chrysogenum*. An amidolyase subunit with a binding site for IPN and a binding site for 6-APA may act cooperatively with an acylCoA:APA acyltransferase subunit that contains a binding site for activated-acyl derivatives and a binding site for 6-APA. In this case, protein-protein interaction would be required for effective function of either subunit. In *E. coli*, this organization of binding sites may occur in separate domains of a bifunctional 40-kDa protein.

Figure 1 Biosynthetic pathways to the sulfur-containing β-lactam antibiotics penicillin G, cephalosporin C, and cephamycin C. Where gene designations are shown in bold print, the gene has been cloned and expressed in *E. coli*; gene designations indicated in plain print are provisional and predict the nature of genes based on the current understanding of a corresponding gene product. The designation *pen*DE for IPN:acylCoA acyltransferase is in plain print to indicate the concept of a gene encoding a polypeptide with an IPN amidolyase domain and an 6-APA:acylCoA acyltransferase domain; an acyltransferase gene has been cloned (Barredo et al., 1989b; Veenstra et al., 1989), but the domains in the gene product have not been proven.

Alternatively acylCoA:IPN acyltransferase may have only one binding site each for IPN, activated-acyl substrate, and 6-APA, or one or more of the binding sites could be formed upon interaction of the subunits. No data exists to assign individual binding sites to the 10- and 30-kDa proteins. [An early estimate of 30 kDa for MW, of a protein with acylCoA: IPN acyltransferase activity (Alvarez et al., 1987) was probably adversely affected by protein-gel interactions. The 40-kDa heterodimer appears to interact more strongly with sizing gels relative to many marker proteins; high concentrations of ethyl glycol added to the elution buffer used with HPLC gel filtration removes this technical problem (Whiteman et al., 1990).]

Mechanistically, the first step in converting IPN to penicillin G or V may involve formation of an L-alpha-aminoadipyl-enzyme adduct with concomitant production of 6-APA, followed by hydrolysis of the L-alpha-aminoadipyl-enzyme adduct to L-alpha-aminoadipic acid (or a cyclic derivative of L-alpha-aminoadipic acid). Conversion of IPN to 6-APA has been observed with extracts of *P. chrysogenum* (Abraham, unpublished) and catalysis of the reaction has been referred to as IPN amidolyase activity cited in Queener and Neuss, 1982). IPN amidolyase and acylCoA:6-APA acyltransferase activities both co-purified with acylCoA:IPN acyltransferase activity; the purified 40-kDa protein exhibited specific activities: 100,000 pmole/s/mg (acylCoA:6-APA acyltransferase activity); 414 pmole/s/mg (phenylacetyl:IPN acyltransferase activity), and 27 pmol/s/mg (IPN amidolyase activity) (Whiteman et al., 1990). Efficient transfer of the L-alpha-amino-adipyl moiety from IPN to the enzyme could require allosteric factors from the second step (see below) in the conversion of IPN to penicillin G or V.

If IPN amidolyase contains a binding site for 6-APA and the postulated L-alpha-aminoadipyl-enzyme adduct reacted faster with 6-APA than with water, one could miss observing formation [^{3}H] 6-APA by simple hydrolysis of [^{3}H] IPN and yet observe formation of [^{3}H] 6-APA upon transfer of the L-alpha-aminoadipyl moiety to exogenously added 6-APA. In studying the conversion of δ-(L-alpha-aminoadipyl)-L-cysteinyl-D-[2-^{3}H] valine to [^{3}H] IPN by cell-free extracts of *P. chrysogenum*, formation of [^{3}H] 6-APA was not reported, but when δ-(L-alpha-aminoadipyl)-L-cysteinyl-D-[2-^{3}H] valine and unlabeled 6-APA were incubated in the extract. [^{3}H] 6-APA was formed (Meesschaert et al., 1980); the extract appeared to form [^{3}H] IPN when incubated with unlabeled ACV and [^{3}H] 6-APA. If this concept of IPN amidolyase is correct, then assignment of IPN amidolyase activity to one or more of the 10-, 30-, or 40-kDa proteins may be easier using the IPN:6-APA acyltransferase activity of the subunit or heterodimer rather than its IPN:H$_2$O acyltransferase activity.

The second step in the IPN:acylCoA acyltransferase reaction is better understood. An acylCoA:6-APA acyltransferase activity catalyzes the transfer of the acyl moeity of PACoA or PoACoA to 6-APA to form penicillin G or V, respectively. Presumably the reaction proceeds with the stoichiometric formation of free coenzyme A (CoAH). No attempts to isolate, identify, and measure CoAH from reaction mixtures have been reported. Queener and Neuss (1982) suggested that acylCoA:6-APA acyltransferase activity involves an acylCoA binding site that interacts primarily with the acyl moiety of acylCoA substrates, and that this site was capable of also binding penicillins via its acyl side chain and that an enzyme-acyl intermediate is formed during the IPN:acylCoA acyltransferase reaction. Many diverse observations recorded in the literature for acylases in the canteloupe line of *P. chrysogenum* strains are easily explainable via this relatively simple model (Queener and Neuss, 1982).

Two independent purifications of *P. chrysogenum* acylCoA:IPN acyltransferase (Alvarez et al., 1987; Whiteman et al., 1990) both utilized the acylCoA:IPN acyltransferase activity as the primary assay for isolating acylCoA:IPN acyltransferase. Like the Oxford group (see above), the Leon group observed that the specific acylCoA:IPN acyltransferase activity (4354 pmol/s/mg) was substantially higher than the specific acylCoA:IPN acyltransferase activity (310 pmol/s/mg) (Alvarez et al., 1987). The high ratio of 6-APA acyltransferase activity/IPN acyltransferase activity can be interpreted in several ways. Formation or hydrolysis of an L-alpha-aminoadipyl-enzyme adduct could be rate limiting. Alternatively, an IPN amidolyase domain of a bifunctional subunit could have been partially inactivated during the purifications or an IPN amidolyase subunit could have been partially inactivated.

The concept of an activated-acyl binding site that reacts covalently with the acyl moiety of PACoA, penicillin G, PoACoA, or penicillin V to form an acyl-enzyme adduct and CoAH or 6-APA predicts that acylCoA: 6-APA acyltransferase should catalyze isotopic exchange ^{35}S between 6-APA and penicillin G or V. In addition, an enzyme catalyzing the latter reaction should also catalyze isotopic exchange of ^{35}S between penicillin G and penicillin V:

$$[^{35}S] \text{ penicillin V} + 6\text{-APA} \rightarrow \text{penicillin V} + [^{35}S]\text{ 6-APA}$$
$$\text{penicillin G} + [^{35}S]\text{ 6-APA} \rightarrow [^{35}S]\text{ penicillin G} + 6\text{-APA}$$

$$[^{35}S] \text{ penicillin V} + \text{penicillin G} \rightarrow [^{35}S]\text{ penicillin G} + \text{penicillin V}$$

In the biosynthesis of cephalosporin C or cephamycin C, the remaining steps differ. Transacetylation of DAOC in *A. chrysogenum* gives cepha-Exogenous 6-APA would stimulate the isotopic exchange of ^{35}S between penicillin G and penicillin V, provided there was neglible hydrolysis of the acyl enzyme. Without exogenous 6-APA, the amount of 6-APA available to drive the exchange would be very small, i.e., equal to the amount of acyl enzyme.

Preuss and Johnson (1967) partially purified an acyltransferase from *P. chrysogenum* that catalyzed the isotopic exchange of ^{35}S between 6-APA and penicillin G or V and the isotopic exchange of ^{35}S between penicillin G and penicillin V. Exogenous 6-APA stimulated the latter exchange.

The enzyme purified by Preuss and Johnson on the basis of trans-acylase activity is probably the same enzyme purified by Alvarez and co-workers and by Whiteman and colleagues on the basis of acylCoA:6-APA acyltransferase activity. Comparing the enzymes isolated by Preuss and Johnson to that isolated by Alvarez and co-workers: Each exhibited activity that was strongly stimulated by thiols such as beta-mercaptoethanol, each did not hydrolyze penicillin G to 6-APA, and each exhibited a pH optimum of ca. 8. Alvarez et al. (1987) observed that acylCoA:6-APA acyltransferase did not form penicillin G from penicillin N and PACoA; likewise, Preuss and Johnson observed that their acyltransferase did not catalyze the isotopic exchange of ^{35}S between 6-APA and penicillin N. The enzyme purified by Preuss and Johnson occurred at elevated levels in *P. chrysogenum* mutants able to produce more penicillin; acylCoA: 6-APA acyltransferase purified by Alavarez and colleagues was absent in a mutant able to make ACV but unable to make penicillin G.

In future studies, the 10-, 30-, and 40-kDa proteins each must be further studied separately and in mixtures. Substrate binding studies with the 10- and 30- kDa subunits, separate and mixed, are needed. Finally, the nature of the L-alpha-aminoadipyl product formed by removal of the side chain from IPN during formation of penicillin G or V needs to be established unequivocally.

C. Biosynthesis of Cephalosporins

Biosynthesis of the bicyclic cephem ring system that characterizes all cephalosporin antibiotics involves two steps: isomerization of IPN to penicillin N and expansion of the thiazoline ring of penicillin N to the dihydrothiazine ring of deacetoxycephalosporin C (DAOC). In the biosynthesis of cephalosporin C by *A. chrysogenum* and the biosynthesis of cephamycin C by *S. clavuligerus* and *N. lactamdurans*, DAOC is hydroxylated to deacetylcephalosporin C (DAC) (Figure 1).

losporin C; transcarbamylation, 7-hydroxylation, and 7-O-transmethylation gives cephamycin C in *S. clavuligerus* and *N. lactamdurans* (Figure 1).

Many other natural cephalosporins, produced by a variety of microorganisms, have been observed. Cephalosporins with moieties other than acetyl or carbamoyl attached to the 3' oxygen of DAC are common in prokaryotes and are probably derived by transferases or, in some cases, by nonenzymatic displacement of acetyl or carbamoyl moieties. A cephalosporin with a 3' aldehyde from a fungus was probably derived by oxidation of DAC. Cephalosporins with glutaryl side chains attached to the 7-amino-cephem nucleus of cephalosporins have been found. They probably were formed by oxidative decarboxylation of D-alpha aminoadipyl side chains.

A complete listing of natural cephalosporins is beyond the scope of this chapter. Biosynthetic and molecular genetic studies have focused on cephalosporin C and cephamycin C. Both of these cephalosporins are used to manufacture semisynthetic cephalosporins with medical utility.

1. Epimerization of a Chiral Center in IPN to Form Penicillin N

IPN is enzymatically isomerized to penicillin N in microorganisms that produce cephalosporins (e.g., *A. chrysogenum* [Jayatilake et al., 1981] and 7-alpha-methoxycephalosporins (e.g., *S. clavuligerus* [Jensen et al., 1983]). The L-alpha-aminoadipyl side chain of IPN is epimerized into the D-alpha-aminoadipyl side chain of penicillin N. The enzyme was named *isopenicillin N epimerase* (Jayatilake et al., 1981; Abraham et al., 1981). IPN epimerase has not been found in producers of penicillin G and V (e.g., *P. chrysogenum, Aspergillus nidulans*).

Recently, the enzyme from *S. clavuligerus*—purified to homogeneity—was observed to isomerize either IPN or penicillin N and to establish an equilibrium between approximately equimolar amounts of the two penicillins (Usui and Yu, 1989). This represents a "racemization of the side chain." However, in 1978, NCIUB listed 11 examples of "racemases" acting on amino acids and derivatives. Each interconverted optical isomers (NCIUB, 1978). Both examples of "epimerases" interconverted diastereomers by epimerizing one chiral center in molecules with more than one chiral center. The systematic name for both epimerases included the designation *x-epimerase*, where *x* indicated the carbon at which epimerization occurred. Since IPN and penicillin N are diastereomers and not optical isomers, the name *IPN epimerase* and not racemase follows precedent. If one uses a numbering system analogous to that of Neuss et al. (1971) for the alpha-aminoadipyl side chain in cephalosporin C (side-chain atoms, amide N —→ carboxyl C, follow ring atoms in numbering), then IPN epimerase would have the systematic designation, *IPN 13-epimerase.*

The IPN epimerase from *S. clavuligerus* is monomeric, has a relative molecular weight of ca. 47,000-50,000, and contains one mole of pyridoxal-5-phosphate per mole of enzyme (Usui and Yu, 1989). The IPN epimerase of *N. lactamdurans* has a molecular weight of 59,000 and contains also pyridoxal-5-phosphate bound to the enzyme (Láiz et al., 1990).

2. Oxidative Ring Expansion of Penicillin N to DAOC

The conversion of penicillin N to deacetoxycephalosporin C (DAOC) is catalyzed by an enzyme activity that has been studied in *A. chrysogenum* (Kupka et al., 1983; Dotzlaf and Yeh, 1987), *S. clavuligerus* (Jensen et al., 1985) and *N. lactamdurans* (Cortes et al., 1987). The enzymes catalyzing the reaction each require Fe^{2+} and O_2 to remove two hydrogen atoms from penicillin N in forming DAOC. Thus the enzymes are oxidoreductases. Each of the enzymes appears to require 2-oxoglutarate as a substrate for activity.

Penicillin N, 2-oxoglutarate oxidoreductase (ring expanding) may be a suitable systematic name for the activity, but is cumbersome. The enzyme activity has been referred to by the trivial designations *ring expanding enzyme, expandase, DAOC synthetase,* and *DAOC synthase* activity. However, *synthetase* implies a *ligase* type mechanism and *synthase* implies a *lyase* mechanism. A "synthetase" mechanism cannot be ruled out for the enzyme from the eukaryote *A. chrysogenum* since ATP stimulates the activity. However, the stimulation is modest; a direct role for ATP in the reaction certainly is not proven and probably is unlikely. ATP does not stimulate the activity in the procaryotes *S. clavuligerus* and *N. lactamdurans.* There is not net oxygenation of substrate in the ring-expansion reaction, but if oxygenation of penicillin N occurred as one phase of ring expansion and deoxygenation by the elimination of the elements of water occurred in a second phase, then a name implying a "lyase" mechanism might be appropriate. However, there is no evidence to support such a mechanism.

Several oxidoreductases — proline hydroxylase, thymidine 2'-hydroxylase, lysine hydroxylase, and thymine 7-hydroxylase — also require 2-oxoglutarate as a cosubstrate. These enzymes are dioxygenases that utilize Fe^{2+} as a cofactor, add one oxygen atom from O_2 to the substrate, and use the other atom for coupled-oxidative conversion of 2-oxoglutarate to succinate and CO_2 (Gunsalus et al., 1975). Succinate is a product of ring expansion in *A. chrysogenum* (Baldwin, unpublished). If ring expansion proceeds, in part, by a decarboxylation-coupled oxidative process paralleling the mechanism of the above hydroxylases, then 2-oxoglutarate utilization should be stoichiometric with formation of succinate and DAOC; furthermore, succinate formed in the presence of ^{18}O-dioxygen should contain an atom of ^{18}O. Although an oxygenated form of penicillin N may be formed as an intermediate in ring expansion, such an intermediate would probably be too labile to isolate; the name *dioxygenase* should probably not be applied to the enzymes that convert penicillin N to DAOC, because there is no net oxygenation of penicillin N in the formation of DAOC. In this review we use the trivial designation *penicillin N expandase* or simply *expandase*. The name is convenient, does not imply an unproven mechanism, and unambiguously designates the activity.

3. Bifunctional and Monofunctional Penicillin N Expandases

The protein catalyzing ring-expansion in *A. chrysogenum* is bifunctional, i.e., it also catalyzes the penultimate step in the biosynthesis of cephalosporin C, the 3'-hydroxylation of DAOC (Figure 1). Scheidegger et al. (1984) partially purified the protein and found that it always exhibited a second activity, catalysis of the hydroxylation of DAOC to form deacetylcephalsporin C (DAC). This "DAOC 3'-hydroxylase" activity remained associated in constant ratio with penicillin N expandase activity. The *A. chrysogenum* protein exhibiting both penicillin-N expandase and DAOC 3'-hydroxylase activities was recently purified to homogeneity (Dotzlaf and Yeh, 1987) and was shown to have a relative molecular mass of 31,000 (Dotzlaf and Yeh, 1987). The bifunctional nature of the enzyme was soon confirmed by an independent purification (Baldwin et al., 1987a; Baldwin et al., 1987b.) As discussed in more detail below, Samson et al. (1987b) cloned the gene corresponding to the pure protein and expressed the open reading frame of that gene in *E. coli*; both penicillin N expandase and DAOC 3'-hydroxylase activities were observed in the recombinant *E. coli*.

Various trivial designations have been used for the bifunctional *A. chrysogenum* enzyme: DAOC synthetase/DAC synthetase, DAOC synthetase/hydroxylase, and expandase/hydroxylase. In this review we use the trivial name *penicillin N expandase/DAOC 3'-hydroxylase* or simply *expandase/hydroxylase*. The name avoids unwarranted mechanistic implications and unambiguously designates the enzyme.

The prokaryotic penicillin N expandases from *S. clavuligerus* and *N. lactamdurans* are smaller than expandase/hydroxylase (M_r 29,500 and 27,000, respectively) and they are monofunctional, i.e., exhibited negligible catalaysis of 3' hydroxylation of DAOC (Jensen et al., 1985; Cortes et al., 1987). Like the eukaryotic expandase/hydroxylase, the two prokaryotic expandases are cytoplasmic monomeric proteins. Using large amounts of pure *S. cluvuligerus* penicillin N expandase of high specific activity, trace hydroxylase activity (< 1% relative to expandase activity) was, in fact, observed [Dotzlaf and Yeh, unpublished]; likewise, using large amounts of pure *S. clavulierus* DAOC 3'-hydroxylase, trace expandase activity (<1% relative to hydroxylase activity) was observed [Baker, manuscript in preparation].

Expandase activity in *A. chrysogenum* is strongly stimulated by ascorbate; ascorbate has a lesser effect on the activity in *S. clavuligerus* and no effect in *N. lactamdurans*. DTT strongly stimulates expandase activity in the fungus and stabilizes the activity in the two prokaryotes.

Purified expandase/hydroxylase from *A. chrysogenum* expands the beta-lactam ring of penicillin N but does not accept IPN, penicillin G, or 6-APA (Dotzlaf and Yeh, 1987). Thus, the enzyme exhibits a high specificity for the D-alpha-aminoadipyl side chain attached to the 6-APA nucleus of its substrate penicillin N.

Dotzlaf and Yeh (1989) have purified to homogeneity the *S. clavuligerus* expandase from the natural host and from an *E. coli* strain that expressed the cloned *S. clavuligerus cefE* gene (Kovacevic et al., 1989, see below).

4. 3'-Hydroxylation of DAOC to DAC

In *S. clavuligerus* (Jensen et al., 1985), the enzyme catalyzing the hydroxylation of DAOC to DAC is separable from penicillin N expandase. In this organism the reaction represents a step in the biosynthesis of cephamycin C, a 7-alpha-methoxycephalosporin (Figure 1). In this review the trivial name *DAOC 3' hydroxylase*, or simply *3' hydroxylase*, is used for this hydroxylating enzyme.

The 3'-hydroxylase activity of *A. chrysogenum* expandase/hydroxylase and the DAOC 3' hydroxylases of *S. clavuligerus* both require Fe^{2+}, 2-oxoglurate, and O_2 to incorporate an atom of oxygen into DAOC (Brewer et al., 1977; Turner et al., 1978; Dotzlaf and Yeh, 1987). The oxygen atom attached at the C3' methylene moiety in cephalosporin C (Figure 1), which is introduced in the conversion of DAOC to DAC, is derived from molecular oxygen (Stevens et al., 1975). Thus it is very likely that DAOC 3' hydroxylase is a decarboxylation-coupled dioxygenase. *Deacetoxycephalosporin C, 2-oxoglutarate 3'-dioxgenase* and *deacetoxycephalosporin C, 2-oxoglutarate:oxygen oxidoreductase* (3'-hydroxylating) would be names for DAOC 3'-hydroxylase activity that appear to conform to the formats of the "recommended" and "systematic" names, respectively, outlined for decarboxylation-coupled dioxygenase enzyme activities by the Nomenclature Committee of the International Union of Biochemistry. However, demonstration that succinate and carbon dioxide are reaction products and ^{18}O incorporation studies are needed to prove a "decarboxylation-coupled dioxygenase" mechanism for 3'-hydroxylation of DAOC.

Dotzlaf and Yeh (1987) observed a slight stimulation of 3'-hydroxylase activity of the *A. chrysogenum* expandase/hydroxylase by ATP; Turner et al. (1978) observed inhibition with a partially purified preparation under slightly different conditions. No stimulation was observed with the *S. clavuligerus* DAOC 3'-hydroxylase (Dotzlaf and Yeh, unpublished).

The 3'-hydroxylase activity of the *A. chrysogenum* expandase/hydroxylase is strongly stimulated by ascorbate and dithireitol (DTT); The *S. clavuligerus* DAOC 3'-hydroxylase is stabilized by DTT, and the activity is weakly stimulated by ascorbate (Dofzlaf and Yeh, unpublished).

The DAOC 3'-hydroxylase activity of the *A. chrysogenum* expandase/hydroxylase and DAOC 3'-hydroxylase from *S. clavuligerus* are both highly specific for DAOC. Substitution of a formyl, phenylacetyl, phenoxyacetyl, glutaramido, or hydrogen for the D-alpha-aminoadipyl side chain of DAOC give compounds that will not serve as substrates (Brewer et al., 1977; Turner et al., 1978).

5. Transcetylation and Transcarbamylation of DAC in the Pathways to Cephalosporin C and Cephamycin C

Cephalosporin C is formed from deacetylcephalosporin C by transfer of an acetyl group from acetylCoA (Figure 1) in *A. chrysogenum*. Mutants have been isolated that lack acetyl-CoA:deacetylcephalosporin C acetyltransferase and accumulate deacetylcephalosporin C (Fujisawa et al., 1973; Fujisawa et al., 1975a). One millimole of Mg^{2+} stimulates the reaction (Fujisawa and Kanzaki, 1975). The enzyme, purified 104-fold (Liersch et al., 1976), is specific for 3-hydroxymethyl-3-em compounds; the absence of or substitution for the D-alpha-aminoadipyl side chain has little effect on the reaction rate, but the enzyme does not catalyze acetyltransfer to DL-serine, DL-homoserine, or the lactone of DAC. The absence of a D-alpha-aminoadipyl-binding site in the enzyme markedly contrasts with the other enzymes in the cephalosporin C pathway. Perhaps this enzyme evolved late and separately from other enzymes in the pathway.

Transfer of the carbamoyl group is involved in the biosynthesis of cephamycin C. Transfer of the carbamoyl group from carbamoylphosphate to DAC to form O-carbamoyl-deacetylcephalosporin C (OCDAC) (Figure 1) has been achieved in cell-free extracts of *S. clavuligerus* (Brewer et al., 1980). The transcarbamylase also accepts 7-alpha-methoxy-deacetylcephalosporin C as substrate. The latter compound can be derived by degradation of cephamycin C.

6. Methoxylation of OCDAC to Form Cephamycin C

The C-7 methoxy group of cephamycins is derived from molecular oxygen and methionine (O'Sullivan and Abraham, 1980). Cell-free extracts of *S. clavuligerus* convert OCDAC but not DAC into 7-alpha-methoxy derivatives in the presence of Fe^{2+}, 2-oxooglutarate, S-adenosylmethionine (SAM) and ascorbate (O'Sullivan and Abraham, 1980). This suggests that 7-alpha-hydroxylation and transmethylation occur after carbamylation (Figure 1).

II. BIOCHEMICAL GENETICS OF PENICILLIN AND CEPHALOSPORIN BIOSYNTHESIS

A. Genetics for Biosynthesis of Precursable Penicillins

1. *P. chrysogenum* Mutants Impaired in Biosynthesis of Precursable Penicillins

Progress in the genetics of the biosynthesis of precursable penicillins, e.g., penicillin G and V, has been slow. In initial studies Sermonti and co-workers isolated nine mutants of *P. chrysogenum* that were impaired in precursable penicillin production (Sermonti, 1956; Cagloti and Sermonti, 1956). Genetic complementation tests were used to indicate whether impaired mutants were altered at the same or at different loci. Complementation analysis with heterozygous diploids provided the initial evidence for the involvement of at least two genetic loci in penicillin biosynthesis (Sermonti, 1959). Biochemical studies of impaired mutants have provided, in a few cases, unequivocal evidence of the steps involved in the biosynthesis of beta-lactam antibiotics. Nash et al., (1974) observed the accumulation of different intermediates of the penicillin biosynthetic pathway by blocked mutants of *P. chrysogenum*. Further studies on the genetics of penicillin biosynthesis involved the characterization of 12 mutants of *P. chrysogenum* producing 10% or less penicillin than their parental strain (Normansell et al., 1979). Analysis of heterozygous diploids formed between them revealed the existence of at least five complementation groups with respect to penicillin production, designated V, W, X, Y, and Z. Mutants of the groups V and W were able to synthesize the tripeptide ACV, whereas mutants of the groups X, Y, and Z failed to form this intermediate.

If each one of the three complementation groups corresponded to mutations in a different gene, then at least three genes would be involved in coding for the synthesis of a tripeptide ACV and for regulating ACV synthesis. Recent data (see Section I.A.1) suggest that only one enzyme is involved in ACV synthesis.

One mutant of group X not only lacked the ability to produce ACV, but also contained reduced levels of acyl-CoA:6-APA acyltransferase (Normansell et al., 1979) assayed by penicillin side-chain exchange (Pang et al., 1984). The lack of ACV synthetase activity and the low level of acyl-CoA:6-APA acyltransferase in this mutant indicate that it may be altered in a regulatory gene controlling the expression of several structural genes.

Two of such regulatory genes (*npe2, npe3*) have been found in a recent study of blocked mutants of *P. chrysogenum* (Cantoral and Martin, unpublished results).

Mutants of groups V and W appear to produce the tripeptide ACV but not penicillin G and could be deficient in ACV cyclase, IPN amidolyase, or acylCoA:6-APA acyltransferase activities (Figure 1). Extracts of mutants of group V could not catalyze penicillin acyl-exchange and therefore are probably deficient in acyl-CoA:6-APA acyltransferase activity. One member of group W possessed almost full penicillin acyl-exchange activity and may, therefore, be blocked between ACV and IPN or between IPN and 6-APA.

Of the 12 nonproducing mutants isolated by Normansell and co-workers, most of them (seven) mapped in the second and third linkage groups (Makins et al., 1983).

2. Genetics of Penicillin G Biosynthesis in *Aspergillus nidulans*

Some strains of *A. nidulans* produce very low levels of precursable penicillins, e.g., penicillin G. Many wild-type *A. nidulans* strains do not form detectable levels of the antibiotic. However, *A. nidulans* has the advantage of a well-developed genetic map. Despite the low level of penicillin production, mutants of *A. nidulans* impaired in penicillin G production have been isolated (Edwards et al., 1974). As in *P. chrysogenum*, over two thirds of the mutants isolated were members of a single complementation group (class A) (Edwards et al., 1974). Several of the wild-type isolates that lacked penicillin biosynthesis were altered in this same locus *npeA* (Cole et al., 1976). The *npeA* gene was located on linkage group VI (Holt et al., 1976). This gene probably controls the formation of ACV, since strains bearing mutations at this locus recovered penicillin G production when supplemented with ACV (Makins et al., 1981; MacDonald et al., 1983). In this respect the *npeA* locus in *A. nidulans* is probably analogous to the *npeY* locus in *P. chrysogenum*. The three other mutations—*npeB, npeC,* and *npeD* were located on three different linkage groups, respectively, and were positioned relative to other loci on chromosomes III, IV, and II, as indicated in Figure 2. Intergeneric cosynthesis of penicillin between impaired mutants of *A. nidulans* and *P. chrysogenum* was reported by Makins et al. (1981) using osmotically fragile mycelia (Makins et al., 1980). Intergeneric complementation suggested that mutations at *npeA* or *npeY* both blocked ACV synthesis. Strains of *A. nidulans* bearing mutations at the *npeC* locus are probably similar to strains of *P. chrysogenum* mutated at the *npeV* locus lacking the ability to produce acyltransferase.

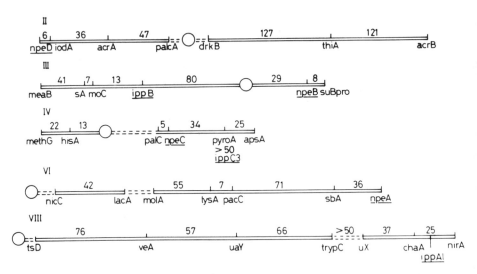

Figure 2 Genetic map of *Aspergillus nidulans*. Only some relevent genetic markers are shown. Loci associated with negation of penicillin synthesis (*npe*) and loci associated with improved production of penicillin (*ipp*) are underlined. (Modified from Simpson and Caten, 1980).

3. Genetic Determinants for Increased Penicillin Production

Some of the mutations leading to increased penicillin production may duplicate the genes or alter the activities of penicillin biosynthetic enzymes. However, many other mutations will probably affect intermediary metabolism, precursor formation, or general regulatory mechanisms (e.g., carbon catabolite or nitrogen metabolism regulation), which affect directly or indirectly penicillin or cephalosporin biosynthesis. Three mutations that caused an increased yield of penicillin in *A. nidulans* designated *pen-A*, *-B*, and *-C*, were positioned after parasexual haploidization studies to chromosomes VIII, III, and IV, respectively (Ditchburn et al., 1976). By sexual crosses, these mutations were mapped as indicated in Figure 2. The nature of these mutations is unknown. It is likely that many different genes influence penicillin biosynthesis. Sexual hybridization between divergent lines of *A. nidulans* with high production penicillin titer has been used for selective increases of penicillin production. However, complementary segregants with substantially improved titers were not found (Simpson and Caten, 1980; Simpson and Caten, 1981).

Since designations for these mutants have not been cited broadly, consideration of changing the designation for the three *Aspergillus* alleles responsible for improved penicillin production to *ippA, ippB*, and *ippC*, respectively, might be considered. The designation *pen* has been suggested for use in naming penicillin and cephalosporin biosynthetic genes in a scheme that would be consistent with precedents set for other "branched" pathways: *pcb* for genes involved in steps shared by both penicillin and cephalosporin biosynthesis (L-alpha-aminoadipic acid, L-cysteine, L-valine to IPN); *pen* for genes involved in converting IPN into precursable penicillins, penicillin G, or penicillin V; and *cef* for genes involved in converting IPN); to end-product cephalosporins, e.g., cephalosporin C and *cmc* for genes involved in converting DAC to cephamycins, e.g., cephamycin C (see Figure 1).

B. Mutants Impaired in Cephalosporin Biosynthesis

The use of blocked mutants that are impaired in cephalosporin biosynthesis was of great importance in deciding whether penicillin N was a precursor of cephalosporin C or whether these two beta-lactam antibiotics were each end products of two pathways branching from a common intermediate. Mutants of *A. chrysogenum* unable to synthesize cephalosporin C or penicillin N were first studied by Lemke and Nash (1971). These mutants were divided into two classes: one class produced a peptide-containing alpha-aminoadipic acid, cysteine, and valine, whereas the other did not form the peptide material. These two classes were complementary, i.e., heterokaryons formed from a peptide-producing mutant and a peptide-negative mutant-produced antibiotic (Nash et al., 1974).

In a different laboratory, three blocked mutants of *A. chrysogenum* (N-2, N-31, and N-79) were found that accumulate the dimer of ACV and also the S-methylthio derivative of the tripeptide (Shirafuji et al., 1979; Shirafuji and Yoneda, 1982). The mutant N2 appeared to be deficient in the activities of three biosynthetic enzyme activities of the pathway: ACV cyclase, IPN epimerase, and penicillin N expandase (Ramos et al., 1986). These results explain why the mutant accumulates ACV and suggest that this mutant is defective in a gene that controls the formation of the three biosynthetic enzymes. Interestingly, however, cephalosporin biosynthesis in mutant N2 could be restored via DNA-mediated complementation using plasmid pPS26, which contains the 1-kb ACV cyclase gene of *A. chrysogenum*, as well as about 3 kb of *A. chrysogenum* DNA flanking the ACV cyclase gene (Chapman et al., 1987; Ramsden et al., 1989). (Similar results have been obtained in an independent study of the mutant N2 with the IPNS gene of *P. chrysogenum* [B. Diez and J.F. Martin, unpublished].) A single base pair change in *pcbC* occurs in N2 and causes an amino acid change (pco → leu); the altered enzyme is inactive (Ramsden et al., 1989).

Yoshida et al. (1978) classified five mutants of *A. chrysogenum* blocked in cephalosporin-C biosynthesis. Type A mutants did not produce penicillins or cephalosporins. Type B mutants formed penicillin N but not cephalosporin C. Cell-free extracts of type A converted penicillin N to a cephalosporin; extracts of type B mutants did not.

Fujisawa and co-workers characterized two distinct types of *A. chrysogenum* mutants that did not accumulate cephalosporin C (Fujisawa et al., 1973, 1975a, 1975b). Some mutants accumulated DAC. The mutation in these cells apparently inactivated the biosynthetic gene encoding the acetyltransferase that converts DAC into cephalosporin C (Fujisawa et al., 1973, 1975a). A different mutant was, however, able to hydrolyze cephalosporin C to form DAC by the action of cephalosporin C acetyl-hydrolase (Fujisawa et al., 1975b).

Queener et al. (1974) isolated a cephalosporin C (CPC)-negative mutant that accumulated more penicillin N and DAOC than the CPC-producing strain from which it was derived. The mutant produced only a trace of DAC, whereas the parent produced significant DAC and cephalosporin C. Total cephalosporins were markedly reduced in the mutant. Thus curtailment of the ability to convert DAOC to DAC in the mutant seemed to be associated with a lower than expected conversion of penicillin N to DAOC.

Liersch et al. (1976) described a mutant that appeared to only produce DAOC, but after testing the products of this strain with more sensitive HPLC systems, the mutant was also found to produce some DAC and CPC (Scheidegger et al., 1984).

For many years it was unclear why mutants with simple blocks in the conversion of DAOC to DAC were not isolated and why mutants with lowered ability to convert DAOC to DAC always seemed to exhibit lower than expected production of DAOC. This can now be explained. In *A. chrysogenum* a single gene codes for a bifunctional protein, penicillin N expandase/DAOC 3'-hydroxylase (Samson et al., 1987b; see discussion in I.C.3 and III.C).

In contrast to the wealth of information on blocked mutants in cephalosporin biosynthesis, not much is known about the genetics of cephalosporin overproduction, although a number of overproducing mutants have been studied from the biochemical point of view (Ramos et al., 1986; Perez-Martinez and Peberdy, 1985; Komatsu et al., 1975; Queener et al., 1984; Shen et al., 1986). This is due to the difficulty of utilizing parasexual genetic analysis in *A. chrysogenum* (Nuesch et al., 1973; Hamlyn and Ball, 1979). Molecular genetic techniques can now partially solve this problem.

III. CLONING AND CHARACTERIZATION OF GENES INVOLVED IN PENICILLIN AND CEPHALOSPORIN BIOSYNTHESIS

The molecular genetics of penicillin- and cephalosporin-producing microorganisms are in an early state of development. Rational construction of one *Acremonium chrysogenum* strain with an extra expandase/hydroxylase gene that allows better conversion of penicillin N to cephalosporin C has been reported (see below). More examples of improved productivity with multiple copies of penicillin and cephalosporin biosynthetic genes can be anticipated. Fusing the open reading frames of these genes to strong fungal promoters may increase expression. Improving expression via multiple copies of positive regulatory genes may also be possible. Improved processes may be developed by expressing modified expandase genes in *P. chrysogenum*. These applications depend on: (a) cloning antibiotic biosynthetic or regulatory genes, (b) cloning structural genes with strong promoters, and (c) construction of hybrid genes expressing cloned native and modified genes in penicillin and cephalosporin-producing microorganisms. Progress in each of these three areas is described below. Also see reviews by Martin and Liras (1988) and Ingolia and Queener (1989).

A. Comparison of *pcbC*, Gene Coding for ACV Cyclase, in Different Penicillin and Cephalosporin Producers

Cloning of the ACV cyclase gene from *A. chrysogenum* was made possible by purifying the enzyme from *A. chrysogenum* ATCC 11550 and determining the first 23 amino-terminal amino acids. A set of synthetic oligonucleotides based on the amino-terminal amino acid sequence was prepared and a cosmid genome library was tested for hybridization to the mixed oligonucleotides probes (Samson et al., 1985). The ACV cyclase gene was identified and sequenced. The open reading frame encoded a polypeptide of M_r 38,416, which had a predicted amino acid sequence that matched the experimentally determined sequence. When this open reading frame was inserted into an *E. coli* expression vector and was transformed into *E. coli*, the recombinant strain produced a new protein co-migrating with authentic ACV cyclase as the major protein of the cell (about 20% of the cell protein) (Samson et al., 1985). The predicted amino acid sequence of the cyclase protein begins with methionine and glycine residues, which are not found in the protein isolated from *A. chrysogenum* cells, which suggests that these residues are cleaved posttranslationally (Chapman et al., 1987). The ACV cyclase has been isolated from *E. coli* transformed with the cloned gene from *A. chrysogenum* and purified to homogeneity (Baldwin et al., 1987b). The protein synthesized in *E. coli* has undergone a slightly different N-terminal processing from that observed in the natural fungal host. In mature *A. chrysogenum* the cyclase loses the terminal methionine and glycine residues, whereas the enzymes isolated from *E. coli* loses only the terminal methionine residue. The different processing observed in *E. coli* has no apparent consequence on the biochemical properties of the enzyme. The K_m and kinetics for the conversion of LLD-ACV to isopenicillin are identical. Recombinant ACV cyclase converted analogue substrates into unusual beta-lactam antibiotics in exactly the same way as the fungal protein (Baldwin et al., 1987b). The ACV cyclase gene from *P. chrysogenum* has also been cloned. In this case, Carr et al. (1986) inserted *Sau*3A-digested fragments of the total DNA of *P. chrysogenum* 23X-80-269-37-2 into the *Bam*HI site in EMBL-3 arms, a lambda-phage-derived vector. Later, the same strategy was followed by Barredo et al. (1989a) to clone the ACV cyclase gene of a different strain of *P. chrysogenum* AS-P-78. The libraries were screened using a heterologous hybridization probe based on the nucleotide sequence of the amino-terminal end of the ACV cyclase gene of *A. chrysogenum*. *E. coli* cells transformed with the *P. chrysogenum* ACV cyclase gene contained ACV cyclase activity, whereas untransformed cells were devoid of it. Results of both laboratories indicate that the cloned open reading frame encodes a polypeptide (Figure 5) of an M_r 37,900 which agrees with the biochemically estimated value of 39,000 $\pm$ 1000 for the purified protein for *P. chrysogenum* (Ramos et al., 1985). However there is one amino acid that differs in the ACV cyclase proteins from strain AS-P-78 and the high producer 23X-80-.269-37-2 (Barredo et al., 1989a). More recently the ACV cyclase gene of *Aspergillus nidulans* was identified by heterologous hybridization with a DNA probe corresponding to the *A. chrysogenum* ACV cyclase (Ramon et al., 1987). The open reading frame encoded a 331 amino acid polypeptide M_r of 37,480. Weigel et al. (1988) cloned the ACV cyclase from another strain of *A. nidulans*. The open reading frame encoded a 331 amino acid polypeptide with M_r of 37,612. Both forms of the gene showed extensive homology with *pcbC* genes of other beta-lactam producing fungi (Figures 3 and 4). The open reading frame of the *pcbC* has been overexpressed in *E. coli*, giving ACV cyclase activity (Ramon et al., 1987; Weigel et al., 1988).

The ACV cyclase genes from *S. clavuligerus* (Leskiw et al., 1988) and from *S. lipmanii* (Weigel et al., 1988) were purified using a scheme similar to the one used to isolate the *A. chrysogenum* ACV cyclase gene. The *S. clavuligerus* and *S. lipmanii* ACV cyclase proteins were purified, the amino-terminal amino acid sequences were determined, synthetic probes were prepared based on the derived amino acid sequence, and the ACV cyclase

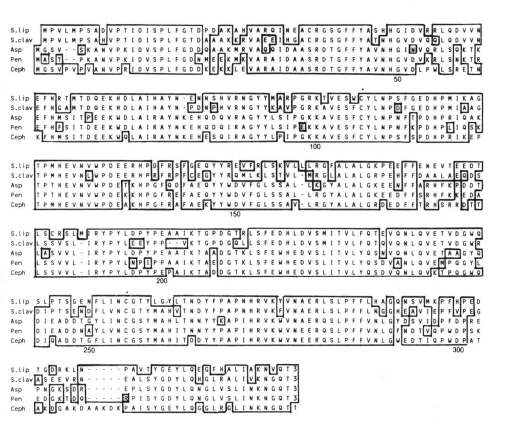

Figure 3 Overall comparison of amino acid sequences of *pcbC* genes. S.lip = *Streptomyces lipmanii* (Weigel et al., 1988), S.clav = *Streptomyces clavuligerus* (Leskiw, 1988), Asp = *Aspergillus nidulans* (Ramon et al., 1988; Weigel et al., 1988), Pen = *Penicillium chrysogenum* (Carr et al., 1986), and Ceph = *Cephalosporium acremonium* (Samson et al., 1985), When two sets of matches occur at a single site (two pairs or a "full house"), all of the matching residues are boxed.

	70.1	56.9	53.9	56.9
79.7		57.0	54.9	55.3
62.5	61.8		78.8	72.6
64.4	64.0	75.7		74.7
64.9	65.4	68.8	72.7	

Figure 4 Pairwise comparison of DNA (lower left) and amino acid (upper right) identities between the *pcbC* genes. S.lip = *Streptomyces lipmanii* (Weigel et al., 1988), S.clav = *Streptomyces clavuligerus* (Leskiw, 1988), Asp = *Aspergillus nidulans* (Ramon et al., 1987; Weigel et al., 1988), Pen = *Penicillium chrysogenum* (Carr et al., 1986), and Ceph = *Cephalosporium acremonium* (Samson et al., 1985).

genes were isolated and characterized. The *Streptomyces* ACV cyclase genes were similar to one another and to the fungal ACV cyclase genes at the DNA sequence and predicted amino acid sequence levels. The degree of similarity of the ACV cyclase genes is summarized in Figure 3. Only identities were counted, and in the case of amino acid comparisons, there were many cases where conservative amino acid changes occurred. An alignment of the predicted amino acid sequences is shown in Figure 3, with amino acid matches enclosed in boxes to indicate the similarities.

The similarity of the ACV cyclase genes suggests an evolutionary relationship among the genes. It has been suggested, based on the fact that the beta-lactam-producing *Streptomyces* species contain the most extensive biosynthetic pathways and that the fungal pathways are truncated versions of the bacterial pathways, that the pathway arose first in *Streptomyces* and was then transferred to fungi (Carr et al., 1986). This same evolutionary relationship was proposed by Ramon et al. (1987) based on the observation that (a) the *C. acremonium* ACV cyclase gene was unusually high in its G/C content relative to other fungi and (b) *Streptomyces* species are known to contain a high percent of G/C in their genome. However, this argument predicts that, if the pathway was transferred to fungi only once, the *Penicillium* and *Aspergillus* ACV cyclase genes should also have a high percent of G/C, but they do not.

The most quantitative argument for a horizontal transfer of the penicillin/cephalosporin biosynthetic pathway from *Streptomyces* to fungi comes from a comparison of the DNA sequences of the genes (Weigel et al., 1988). An estimate for the time of divergence of genes can be obtained from an approximated rate of nucleotide substitution of 1×10^9 nucleotide changes per site per year (Li et al., 1985). Using the ACV cyclase gene as a molecular clock, the evolutionary relationships shown in Figure 5 are obtained. The predicted relationships for the fungal genera are the same using the ACV cyclase or 5S RNA (Hori and Osawa, 1979) as molecular clocks. However, the predicted time of divergence of the bacterial and fungal ACV cyclase genes is only about 370 million years ago, whereas prokaryotes and eukaryotes were supposed to have split about 2 billion years ago. Therefore, the DNA sequence comparisons predict that a primordial penicillin/cephalosporin biosynthetic pathway was transferred about 370 million years ago, probably from a bacterium to a fungal cell.

B. Site-Directed Mutagenesis of an ACV Cyclase Gene

Information about ACV cyclase activity has been obtained through mutagenesis of the cloned genes and expression of the mutant proteins in *E. coli*. Biochemical experiments had shown previously that sulfhydryl-re-

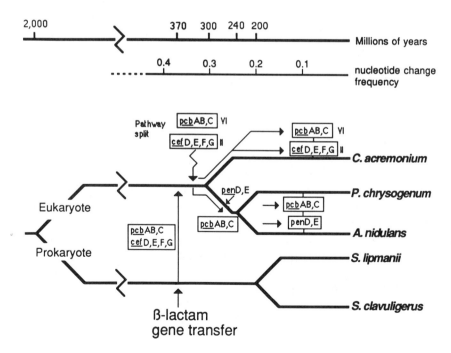

Figure 5 Phylogenetic tree based on DNA sequence identities of the *pcbC* genes.

active alkylating agents inactivated ACV cyclase, suggesting the involvement of a cysteine residue in activity (D. Perry and E. P. Abraham, personal communication). The fungal ACV cyclase genes contain only two cysteine residues, and the location of these residues is conserved in all three (see Figure 3, Cys106 and Cys255, *A. chrysogenum* numbering). The bacterial ACV cyclase genes also contain these same two cysteines and two others. The cysteine residues in the *A. chrysogenum* ACV cyclase were changed to serine residues, one at a time and in combination. After expression in *E. coli*, the ACV cyclase protein with Cys106 mutated to a serine was found to have a specific activity that is only 5% of wild type, and the K_m of the protein was increased about fivefold (Samson et al., 1987a). Mutation of the other cysteine residue, Cys255, had little effect on activity or K_m, either alone or in combination with the Ser106 mutation. The fact that the ACV cyclase had reduced but easily demonstrable activity in the absence of any cysteine residues suggests that a free sulfhydryl is important but not essential for activity.

C. *cefEF A. chrysogenum* Gene Coding for a Bifunctional Enzyme: Penicillin-N Expandase/DAOC 3′-Hydroxylase

Cloning of the structural gene of *A. chrysogenum* for expandase/hydroxylase and expression of the open reading frame in *E. coli* provided unambiguous proof that penicillin N expandase and DAOC 3′-hydroxylase activity are functions of a single polypeptide (Samson et al., 1987b) (Figure 2). The penicillin N expandase and DAOC 3′-hydroxylase activities (Dotzlaf and Yeh, 1987) of *A. chrysogenum* copurified from crude extract to homogenous protein. Sequencing of the purified protein was frustrated by an apparent blockage of the amino terminus of the protein, so the internal amino acid sequence was obtained by cleavage of the enzyme followed by purification and sequencing of peptide fragments. DNA probes were prepared based on parts of the derived amino acid sequence and the corresponding gene was cloned (Samson et al., 1987b). Primary DNA sequence analysis showed that one open reading frame encoded all of the experimentally derived amino acid sequences, and these amino acid sequences accounted for about half of the amino acids encoded by the open reading frame. A protein of molecular weight 36,461 Da was encoded by the open reading frame. When this open reading frame was inserted into an expression vector and expressed in *E. coli*, crude extracts from the *E. coli* cells contained both penicillin expandase and DAOC 3′-hydroxylase activities, unambiguously demonstrating the bifunctional nature of the polypeptide encoded by the cloned open reading frame.

D. *cefE, S. clavuligerus* Gene Coding for Monofunctional Penicillin N Expandase

The bifunctionality of the *cefEF* gene product is particularly interesting when compared to the situation in *Streptomyces* species, where the penicillin-N expandase and DAOC 3'-hydroxylase activities are easily separable by ion-exchange chromatography (Jensen et al., 1985). Recent experiments indicate that the separate expandase in *S. clavuligerus* results from the existence of a gene that codes for expandase but not hydroxylase. (However, see Section C. 3.) Kovacevic et al. (1989) have used an amino-terminal amino acid sequence from purified penicillin N expandase from *S. clavuligerus* to clone the corresponding gene, *cefE*. This gene, which encodes a protein with a predicted molecular weight of 34,633 Da, has been expressed in *E. coli* and leads to the production of strong penicillin N expandase activity and trace, negligible DAOC 3'-hydroxylase activity. Another gene codes for monofunctional 3'-hydroxylase in cephalosporin-producing *Streptomyces*, and that gene has been cloned and expressed in *E. coli* (J.R. Miller, personal communication).

Comparison of the fungal and bacterial genes that code for penicillin N expandase and DAOC 3'-hydroxylase is of interest for several reasons. For example, it should be possible to determine whether or not the fungal expandase/hydroxylase gene was derived by a fusion of separate primordial expandase and hydroxylase genes that evolved first in *Streptomyces*, as predicted by the horizontal transfer hypothesis proposed above. Gene comparisons that include the analysis of the predicted prokaryotic DAOC 3'-hydroxylase gene would help to identify potential active-site regions in the predicted amino acid sequences of today's genes.

Comparison of the expandase/hydroxylase from *A. chrysogenum* and expandase from *S. clavuligerus* is already intriguing. The comparison shows that the prokaryotic expandase gene is homologous to the expandase/hydroxylase gene, but it lacks nucleotides corresponding to the 19 carboxy-terminal amino acids of the *A. chrysogenum* expandase/hydroxylase. Perhaps the expandase/hydroxylase gene in *A. chrysogenum* ATCC 11550 arose by deletion of one of two nearly identical regions of an ancestral fused gene, where that region served the same function in each domain of the fused gene product. Alternative concepts have been suggested (Queener, 1990).

The *A. chrysogenum* penicillin N expandase/DAOC 3'-hydroxylase and the ACV cyclase genes show only limited similarity. For example, the overall percent identity of the predicted amino acid sequences is only about 11%. However, some intriguing similarities do exist. The optimal alignment of the penicillin N expandase/DAOC 3′-hydroxylase and ACV-cyclase-predicted amino acid sequences lines up the Cys106 residue of ACV cyclase (substrate binding was described above) with a corresponding cysteine residue (Cys100) in the penicillin N expandase/DAOC 3′-hydroxylase. Furthermore, a 10 amino acid region from ACV cyclase and an amino acid region from penicillin N expandase/DAOC 3′-hydroxylase, which include these respective cysteine residues, are 50% identical.

E. Gene Coding for IPN:acylCoA Acyltransferase

A purified preparation of acyltransferase that catalyzed the formation of penicillin G from phenylacetylCoA and 6-APA or (with a lower affinity) from IPN, was recently shown to contain three proteins sized at ca. 40, 29, and 10 kDa; the N-terminal sequence of a 29-kDa protein was used to isolate *P. chrysoenum* DNA that contained an open reading frame with three introns (Veenstra et al., 1989; Barredo et al., 1989b). The deduced amino acid sequence of the open reading frame would encode a 39-kDa protein. A DNA sequence in the gene could be found that corresponded to the N-terminal sequence of the 10-kDa protein, and downstream of this sequence another DNA sequence was found that corresponded to the N-terminal sequence of the 29-kDa protein (Veenstra et al., 1989; Barredo et al., 1989b). The large protein, estimated at 40-kDa, in the purified acyltransferase preparation corresponds to a heterodimer formed from the 10- and 29-kDA protein (Whiteman et al., 1990); 10- and 29-kDa subunits are probably formed by proteolysis of the 39-kDa protein encoded in the cloned gene.

A vector bearing a *pyrG* gene and the cloned *P. chrysogenum* gene complemented the *pyrG* mutation in a *P. chrysogenum* strain that also contained an *npe8* mutation. The *npe8* mutation is associated with the absence of acyltransferase activity and inability to produce penicillin. Transformants that were penicillin positive, acylytransferase positive were readily isolated (Veenstra et al., 1989; Barredo et al., 1989b).

The 10-kDa protein and the 29-kDa protein may correspond to proteolytic products formed from splitting apart an IPN amidolyase domain and a 6-APA:acylCoA acyltransferase domain in a 39-kDa IPN:acylCoA acyltransferase. Such speculative concepts should be testably by further biochemical characterization of the three proteins in the purified preparation of Veenstra et al. (1989). These characterizations will significantly elucidate the nature of IPN:acylCoA acyltransferase in *P. chrysogenum*. The work of Veenstra, Barredo and Whiteman, and colleagues represents an important advance towards understanding the biosynthesis of precursable penicillins.

IV. CLONING SYSTEMS AND EXPRESSION VECTORS

Introduction of exogenous DNA into both *P. chrysogenum* and *A. chrysogenum* has been difficult. In order to deliver exogenous DNA into the organism, it is necessary either to render the cell wall permeable to DNA or to remove it by producing protoplasts. Several laboratories have developed genetic transformation procedures.

A. Transformation of *A. chrysogenum*

A hygromycin B phosphotransferase (HPT) gene from *E. coli* was used to develop a transformation system for *A. chrysogenum* (Queener et al., 1985). The gene was cloned, sequenced, and a hybrid HPT gene was constructed by juxtaposing the bacterial HPT open reading frame next to a promoter from the *Saccharomyces cerevisiae* phosphoglycerate kinase (PGK) gene. The resulting PGKp/HPTorf marker proved to be an effective dominant selectable marker in *S. cerevisiae* (Kaster et al., 1984) and was used successfully to establish low-frequency PEG/CaCl$_2$-mediated transformation of wild-type *A. chrysogenum* protoplasts. Expression of the HPTorf in *A. chrysogenum* gave rise to hygromycin B-resistant transformants among regenerated cells of the *A. chrysogenum* wild-type strain (ATCC 11550). Strain 394-4, a less vigorous derivative of ATCC 11550, previously selected via many intermediate strains for its high yield of cephalosporin C, was transformed only rarely with vectors containing the PGKp/HPTorf marker. Southern hybridization studies showed that vector DNA had integrated into the high molecular weight DNA of recipient strains.

To improve the transformation frequency, a new hybrid HPT gene, IPNSp/HPTorf, was constructed by juxtaposing the HPTorf next to a promoter, IPNSp, from the *A. chrysogenum* ACV cyclase gene, *pcbC*. Vectors containing the IPNSp/HPTorf marker transformed *A. chrysogenum* ATCC 11550 at a frequency ca. 100-fold greater than vectors containing either PGKp/HPTorf or HPTorf with no promoter (Skatrud et al., 1987a). Southern hybridization studies again showed that vector DNA had integrated into the high molecular weight DNA of recipient strains.

To further improve transformation, three approaches were explored. Addition of an *A. chrysogenum* mitochondrial DNA fragment that functioned as an autonomous replication sequence in yeast (Skatrud and Queener, 1984) to vectors containing the IPNSp/HPTorf did not improve the transformation frequency. Likewise, the addition of a fragment of *A. chrysogenum* DNA containing DNA from three ribosomal RNA genes (Carr et al., 1987) did not improve transformation. However, better procedures for producing and handling *A. chrysogenum* protoplasts improved the transformation frequency for ATCC 11550 by an additional two to threefold (Skatrud et al., 1987b). The improved marker and procedures produced transformants of strain 394-4 at an adequate rate.

Next, the ability of this improved system to deliver and express a cloned cephalosporin biosynthetic gene in *A. chrysogenum* was tested. Plasmid pPS26 was constructed; it contained IPNSp/HPTorf and an *A. chrysogenum* DNA fragment that included the *pcbC* gene. pPS26 was used to restore cephalosporin C synthesis in a mutant deficient in ACV cyclase (complementation of this mutant is discussed in detail below). Transformation of wild-type *A. chrysogenum* with the plasmid and screening 100 transformants for improved cephalosporin C synthesis showed that none reproducibly exhibited cephalosporin C titers above the range of cephalosporin C yields exhibited by 100 natural selectants (Chapman et al., 1987) a similar result was later obtained when another plasmid containing the *A. chrysogenum pcbC* gene (but less adjacent *A. chrysogenum* DNA) was used to transform strain 394-4, a strain used to produce cephalosporin C commercially. The results suggested that ACV cyclase was not the rate-limiting step in cephalosporin C biosynthesis in that strain.

To ascertain the likely rate-limiting step in strain 394-4, the amount of pathway intermediates excreted into broth were measured. The amount of penicillin N produced by *A. chrysogenum* strain 394-4, a high producer of cephalosporin C, was shown to be significant (Queener et al., 1988; Skatrud and Queener, 1989). Although uptake is possible in certain *Streptomyces* mutants (Mahro and Demain, 1987), excreted penicillin N does not reenter *A. chrysogenum* cells to a significant degree and is no longer available for conversion to cephalosporin C. The amounts of penicillin N produced by *A. chrysogenum* strain 394-4 suggested that penicillin N expandase might be rate limiting for cephalosporin C biosynthesis in this strain and that cephalosporin C production could be increased by providing more intracellular penicillin N expandase activity to increase the conversion of penicillin N to cephalosporins in that strain.

Strain 394-4 was transformed with a plasmid, pPS56, which contains the IPNSp/HPTorf and an *A. chrysogenum* DNA fragment that includes the penicillin N expandase/DAOC 3'-hydroxylase gene, *cefEF*. One of eight hygromycin B-resistant clones isolated in the first transformation experiment produced approximately 40% more cephalosporin C than its high-producing parent in shake-flask fermentations. The amount of penicillin N was markedly reduced in LU4-79-6. Extracts of LU4-79-6 contained 30-80% more penicillin N expandase activity than its parent, strain 394-4.

In both recipient strain 394-4 and transformant LU4-79-6, eight chromosomes were observed after separation by transverse alternating field electrophoresis. Six were resolved; chromosomes #2 and #3 migrated as a poorly resolved doublet. (Note: in *C. acremonium,* chromosomes have been numbered: smallest (#1) to largest to largest (#8) [Skatrud and Queener, 1989].) In both the recipient and transformant, the endogenous *cefEF* gene was detected (hybridization to a *cefEF* gene probe) in the slower migrating DNA of the doublet. In the transformant, the IPNSp/HPTorf-vector DNA was detected (hybridization to nonfungal vector sequences) in the slower migrating DNA of the doublet (Queener et al., 1988; Skatrud et al., 1989). In a separate experiment, Southern hybridization analysis of enzyme-restricted DNA from the transformant indicated that vector DNA had integrated nonhomologously into the chromosomal DNA of the recipient strain (Queener et al., 1988; Skatrud and Queener, 1989).

An advantage for LU4-79-6 has also been observed in pilot plant vessels (Queener et al., 1988; Skatrud et al., 1989). Large-scale experiments must be conducted to determine the practical utility of this strain. Permission from the National Institutes of Health (USA) to conduct such experiments and to use strain LU4-79-6 has been requested (Foglesong, 1988) and recently granted (Wyngaarden, 1988).

Transformation of *A. chrysogenum* using a G418 resistance marker has been reported (Penalva et al., 1985); however, no *A. chrysogenum* strain-improvement applications have been reported using G418. Efficiency-of-plating experiments have shown hygromycin B to be superior to G418 as an agent for selecting *A. chrysogenum* transformants (Skatrud and Queener, unpublished observations). Hygromycin B kills untransformed *A. chrysogenum* more effectively than G418, and purified hygromycin B is available commercially. The hybrid HPT marker very effectively renders *A. chrysogenum* resistant to hygromycin B. The background of nontransformed colonies on selective plates can be avoided by the proper choice of the hygromycin B concentration, which should be determined for each recipient strain. Highly developed strains used for cephalosporin C production are more sensitive to hygromycin B than wild-type ATCC 11550.

B. Transformation of *Penicillium chrysogenum* by Complementation of Auxotrophs

A PEG-mediated high-frequency transformation system for protoplasts of *P. chrysogenum* has been developed (Martin et al., 1987; Cantoral et al., 1987). The mechanism by which cellular membranes become competent for DNA uptake is not known. PEG is known to induce protoplast fusion; DNA may be internalized along with membrane during this process.

Mutants of *P. chrysogenum* Wis 54-1255 resistant to 5'-fluoroorotic acid (5'-FOA) were observed in a population of cells treated with nitrosoguanidine at a frequency of ca. 1 in 6×10^6 survivors. About 10% of all 5'-FOA-resistant mutants showed dependence on uracil for growth when tested on plates with or without uridine. Some of these auxotrophs did not grow at all in the absence of exogenous uracil, did not revert, and were adequate recipient strains for transformation experiments (Diez et al., 1987).

The *pyrG* auxotrophy of uracil auxotrophs chosen as recipients for transformation was confirmed biochemically. The OMP-decarboxylase activity of the wild-type and uracil mutants was measured using a method based on the conversion of [^{14}C] OMP into [^{14}C] UMP. Both nucleotides were easily separated by thin-layer chromatography on polyethylenimine plates (Diez et al., 1987). The parental strain of *P. chrysogenum* AS-P-78 showed an OMP-decarboxylase activity of 62 mU/mg of protein, whereas mutant strain *pyrG1* lacked OMP-decarboxylase in extracts prepared at different times during culture.

Suitable uracil auxotrophs of both penicillin-producing and nonproducing mutants were used as recipient strains for vector-mediated transformation. Two kinds of transformant colonies were observed (Figure 6): type 1, large colonies that grew normally on selective medium, and type 2, very small colonies that did not develop into mature transformants. When transferred to selective medium, type 1 cells grew normally and type 2 cells failed to grow. Type 1 cells could grow without uracil. Two similar types of transformants were found in the transformation of *A. nidulans* with vectors carrying the acetamidase gene (Tilburn et al., 1984; Wernars et al., 1985). The poorly growing fungal transformants probably experience transient expression of the vector-borne selectable marker.

In type 1 uracil-independent *P. chrysogenum* transformants, the plasmid DNA appeared to be integrated into chromosomal DNA, since (a) no free plasmid was detected by Southern hybridization and (b) the hybridization bands obtained after digestion of the total DNA of the transformants with different restriction enzymes were different from the bands obtained when the plasmid vector was digested with the same enzymes (Cantoral et al., 1987; Diez et al., 1987).

One of the transformant strains studied in detail showed a five-fold higher activity for orotidine-5'-monophosphate decarboxylase (coded by the *pyr4* gene) than did the wild-type strain. This observation indicated that the vector-borne *pyr4* gene was efficiently expressed in *Penicillium* (Diez et al., 1987).

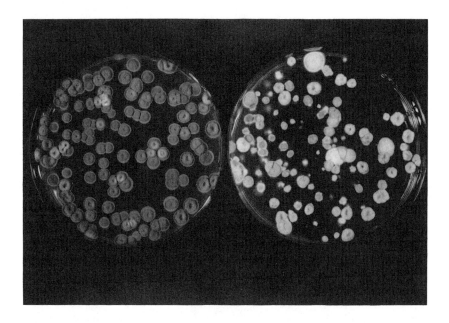

Figure 6 Transformants of *P. chrysogenum* AS-P-78 using plasmid pDJB2. Left: Regeneration of protoplasts of *P. chrysogenum* Wis 54-1255 (Aza^R *pyrGl*) in minimal medium supplemented with uridine (70 μg/ml). Note the homogenous aspect of the colonies. Right: Transformants of *P. chrysogenum* Wis 54-1255 (Aza^R*pyrGl*) without uridine. Note the appearance of large transformant colonies and small abortive colonies. (From Cantoral et al., 1987.)

The efficiency of transformation obtained (3 $\times$ 10^3 transformants/μg of DNA) should be high enough to enable the direct isolation of fungal genes by assaying their expression in *P. chrysogenum*. In fact, an analogous *pyr4*-based transformation system for *A. nidulans*, which exhibits a similar transformation frequency, has been used for the direct cloning of the structural gene for isocitrate lyase (Ballance and Turner, 1985).

An efficient transformation system based on complementation of *trpC* auxotrophs by a vector-borne *P. chrysogenum trpC* gene has recently been reported for *P. chrysogenum* (Picknett et al., 1987; Sanchez et al., 1987). This host-vector system has provided a valuable tool for gene transfer in *Penicillium*, but the transformation frequency has been lower than the *pyrG* system and its usefulness is limited in some cases by the lack of suitable auxotrophic recipient strains. Although some auxotrophs are easy to obtain (Diez et al., 1987), introduction of auxotrophy by mutagenesis can decrease the antibiotic titers of highly productive strains.

C. Dominant Markers for Transformation of *P. chrysogenum*

The acetamidase gene from *A. nidulans* — a dominant marker that does not require auxotrophic recipients — has been used to select for transformed *P. chrysogenum* cells (Beri and Turner, 1987; Penalva et al., 1985; Diez and Martin, unpublished; Skatrud et al., 1987c). Wild-type *P. chrysogenum* cannot use acetamide as its sole nitrogen source, since it lacks acetamidase, an enzyme required to cleave the acetamide to acetate and ammonium. When transformed with plasmid p3SR2 that carries the acetamidase gene, *P. chrysogenum* can grow. The *amdS* transformation system has been used to introduce extra copies of the *P. chrysogenum* ACV cyclase gene, *pcb*C, into a *P. chrysogenum* strain that produced large amounts of penicillin V; some increase in IPNS activity was found in the extracts of three transformants, but none showed an increase in penicillin production (Skatrud et al., 1987c). These results indicated that IPNS was not the rate-limiting step in this strain. The genome of other high-producing strains of *P. chrysogenum* (not a transformant) was shown to already have an amplified unit of DNA, which included the *pcb*C gene (Smith et al., 1988; Barredo et al., 1989c).

Recently, a hybrid gene was constructed by splicing a promoter, IPNSp, from the *pcb*C gene of *P. chrysogenum* to the open reading frame of the expandase gene from *S. clavuligerus*. The resulting hybrid gene was introduced into *P. chrysogenum* using the *amdS* transformation system. The hybrid expandase gene was expressed in *P. chrysogenum*; extracts of the transformant clearly exhibited expandase activity, whereas extracts of the recipient *P. chrysogenum* strain did not (Cantwell et al., 1990).

A second transformation system with a dominant marker for *P. chrysogenum* has been developed using agents to alter the *P. chrysogenum* cell membrane to facilitate hygromycin B uptake. This system employed a hybrid hygromycin B phosphotransferase (HPT) gene containing a promoter from the *P. chrysogenum* ACV cyclase gene spliced to the open reading frame of a bacterial HPT gene (Skatrud and Queener, unpublished).

Transformation systems for *P. chrysogenum* that employ mutant beta-tubulin genes from *Neurospora crassa* and *Penicillium chrysogenum* as selective markers have been developed; the mutant genes expressed in the fungi make them resistant to benomyl (Chiang, 1988). The transformation frequency for these systems is ca. 10 transformants per microgram of DNA. More recently, transformation systems have been developed for *P. chrysogenum* using a gene that provides resistance to phleomycin (Kolar et al., 1988) and using a oligomycin-resistant ATP synthase subunit gene that provides resistance to oligomycin (Bull et al., 1988).

Transformation by dominant markers and by auxotrophic complementation each has advantages and disadvantages. Transformation of *P. chrysogenum* by dominant selectable markers is still very inefficient relative to transformation by the *pyrG* and *trpC* genes. High-efficiency transformation is essential for cloning genes by direct complementation. Gene dosage studies and other direct practical applications of molecular genetic techniques, such as gene disruption or gene replacement, can be carried out in each system, e.g., see the description of the use of *amdS* system for returning ACV cyclase gene to *P. chrysogenum* (above). The *amdS* system has the advantage of not requiring the introduction of an auxotrophic mutation into recipients prior to transformation. The HPT system requires careful adjustments of the concentrations of permeabilizing agents, which is not required by the *amdS, pyrG,* or *trpC* systems.

D. Transformation of *Aspergillus nidulans*

A. nidulans has no commerical importance for antibiotic production. However, it is an interesting model for studying the molecular genetics of precursable penicillin production in fungi, since a detailed genetic map with more than 500 loci is known (Penalva et al., 1985). Two transformation systems for *A. nidulans* were initially reported by Ballance et al. (1983) and Tilburn et al. (1984). The system described by Ballance and co-workers uses the *pyr4* gene of *Neurospora crassa* to complement a mutation in the corresponding *pyrG* gene of *A. nidulans* (see also Chapters 1-4). On the other hand, the system described by Tilburn et al. (1984) uses the cloned acetamidase gene (*amdS*) of *A. nidulans*. Other markers used in vectors to transform *A. nidulans* include the cloned *trpC* (Yelton et al., 1985) and *argB* (John and Peberdy, 1984; Johnstone et al., 1985). Different factors affecting the transformation of *A. nidulans* have been reviewed recently by Johnstone (1985).

E. Autonomously Replicating versus Integrative Vectors

From the industrial point of view, vectors that integrate into the host chromosome are preferred for antibiotic production because the transformants they produce are very stable. When a particular biosynthetic gene is introduced in penicillin- and cephalosporin-producing fungi by an integrative vectors, it has been stably inherited during mitotic division (Skatrud and Queener, 1984; Diez et al., 1987; Diez and Martin, unpublished; Skatrud et al., 1989).

Current integrative transformation systems available for penicillin- and cephalosporin-producing fungi are suitable for a variety of manipulations. As described in Section IV.A, the IPNSp/HPTorf system has been used to construct an *A. chrysogenum* strain containing an extra copy of a biosynthetic gene; the strain exhibited improved cephalosporin C production. As described in Section IV.B, a *pyrG* system has been used to directly clone a fungal gene by complementation of a mutant strain of *A. nidulans*. Recently, integrative transformation has been used for disrupting antibiotic biosynthetic genes in *A. chrysogenum* (Skatrud et al., manuscript in preparation).

A large region of homology between vector and recipient DNA appears to be less important in integrative transformation in *A. chrysogenum* than in *S. cerevisiae*. Certain plasmids that were unable to replicate autonomously in yeast transformed *S. cerevisiae* by integration 100-fold more effectively when *S. cerevisiae* DNA from genes coding for ribosomal RNA (rRNA) was inserted in those plasmids. In *A. chrysogenum*, the addition of fragments of *A. chrysogenum* rRNA genes to integrative vectors did not enhance the transformation frequency (Carr et al., 1987; Skatrud et al., 1987b; see Section IV.A). Integrative transformation appears to occur primarily by homologous recombination in yeast and by heterologous recombination in penicillin-and cephalosporin-producing fungi.

High-copy autonomously replicating vectors will probably not be employed in *P. chrysogenum* or *A. chrysogenum* for the practical production of penicillins and cephalosporins. A number of high-copy autonomously replicating vectors have been developed from a natural plasmid in *S. cerevisiae*. Natural plasmids from *P. chrysogenum* or *A. chrysogenum* have not been reported, and no high-copy autonomously replicating plasmids have been developed for these fungi. Even if such fungal vectors were developed, the relatively low stability of yeast high-copy autonomously replicating vectors indicates that they would still be too unstable for practical use in making antibiotics. Maintaining high-copy plasmids by selection would probably not be amenable to the balanced metabolism necessary for high antibiotic production.

If developed, high-copy autonomously replicating vectors should be more efficient than integrative vectors in the direct isolation of structural or regulatory genes in penicillin- and cephalosporin-producing fungi. In

S. cerevisiae, very high transformation frequencies (up to 10^4 transformants/μg DNA) obtained with autonomously replicating vectors have certainly aided the direct cloning of yeast genes by the complementation of yeast mutants.

Attempts have been made to construct plasmids that would replicate autonomously in *A. chrysogenum* in an effort to increase the frequency of transformation observed with integrative vectors. Early studies showed that a 1.9-kb mitochondrial (mt) DNA sequence from *A. chrysogenum*, when integrated into a bacterial plasmid or a yeast integrative plasmid, could function as an autonomous replication sequence (ars) in *S. cerevisiae* (Tudznski and Esser, 1982; Skatrud and Queener, 1984). However, as mentioned in Section IV.A, the presence or absence of this mt DNA with the dominant selectable marker, IPNS/HPTorf, in *A. chrysogenum* integrative vectors did not affect the frequency of transformation of *A. chrysogenum* by those vectors (Skatrud and Queener, 1984).

There have also been attempts to construct plasmids that would replicate autonomously in *P. chrysogenum*. Two fragments of a mt DNA sequence were introduced into a plasmid carrying a bacterial kanamycin resistance gene. The parent plasmid and the plasmids with the mt DNA were transformed separately into *P. chrysogenum*. Comparisons of the respective transformants (resistant to G418) suggested that the mt DNA fragments enhanced autonomous replication in *P. chrysogenum*. However, transformation frequencies with all of the plasmids were very low (Stahl et al., 1987). Introduction of the *Aspergillus* "*ans1*" sequence into *pyrG* integrative vectors bas been shown to modestly enhance the total efficiency of transformation of *P. chrysogenum* by that system (Diez et al., 1987).

F. Improved Expression from Cloned Promoters

Cloned promoters and other regulatory sequences should be useful in developing vectors capable of efficiently expressing cloned genes in industrial organisms that produce penicillins and cephalosporins. High intracellular concentrations of primary metabolites that serve as precursors of antibiotics occur following the cessation of macromolecular synthesis (i.e., growth) and are required for efficient antibiotic synthesis (Martin and Liras, 1981; Martin and Liras, 1985; Drew and Demain, 1977). The modification of cloned promoters or the replacement of weak promoters with promoters from strongly expressed genes of primary metabolism (e.g., *pgk* promoter) might be used to increase the production of L-alpha-aminoadipic acid, L-cysteine, and L-valine that are required for both penicillin and cephalosporin production and might be used to increase the production of primary metabolite precursors of the methoxy and 3'-C substituents found in many cephalosporins.

Genes involved in converting primary metabolites to antibiotics are expressed only at low growth rates after most vegetative growth has already occurred (Martin and Demain, 1980; Demain et al., 1983). They are controlled by carbon catabolite and nitrogen metabolic regulatory mechanisms, which are only beginning to be understood at the molecular level (Martin and Aharonowitz, 1983; Revilla et al., 1986; Barredo et al., 1988; Drew and Demain, 1977). In the future it may be possible to alter the promoters of these genes so that they do not respond to negative regulatory mechanisms.

It may be possible to substitute promoters from strongly expressed genes involved in primary metabolism (e.g., promoter from *A. chrysogenum PGK* gene) for weaker promoters from penicillin/cephalosporin biosynthetic genes to increase antibiotic production. However, in *Streptomyces* and *Nocardia* this could lead to the production of penicillins and cephalosporins in the growth phase and thereby inhibit growth.

Conversely, promoters from penicillin/cephalosporin biosynthetic genes (e.g., the promoter from the ACV cyclase gene of *A. chrysogenum*) could be used to sustain the production of the primary metabolite precursors required for antibiotic synthesis. These hybrid genes would be added to their natural counterparts.

There has been some work with cloned promoters in *A. chrysogenum*. A promoter (IPNSp) from the *A. chrysogenum* ACV cyclase gene appears to express the HPTorf in a hybrid gene, *IPNSp/HPTorf,* more efficiently than does the promoter from the yeast *PGK* gene in an analogous hybrid gene, *PGKp/HPTorf* (Skatrud et al., 1987a). Preliminary experiments in *A. chrysogenum* indicate that the expression of the HPTorf by a promoter from the *A. chrysogenum* ACV cyclase gene may be stronger than the expression from a promoter from the *A. chrysogenum leu2* gene (John S. Wood, unpublished). It would be interesting to compare the expression of the *IPNSp/HPTorf* to that from a hybrid HPT gene containing the *A. chrysogenum pgk* promoter.

V. FUTURE OUTLOOK

The development of the basic tools of molecular genetics in penicillin- and cephalosporin-producing fungi and *Streptomyces* has provided the groundwork for rapid progress in our understanding of the structural and functional organization of the genes involved in the biosynthesis of penicillin and cephalosporin antibiotics. The ACV cyclase genes from *A. chrysogenum* and *P. chrysogenum* are now available in several laboratories and are being studied by in-vitro mutagenesis. The cloning of ACV cyclase genes from three bacteria have been reported; the expandase/hydroxylase gene from *A. chrysogenum* and the expandase gene from *S. clavuligerus* have been cloned. The gene encoding IPN epimerase has also been cloned. Efficient expression of the open reading frames from all of these genes in *E. coli* has been obtained by Ingolia and colleagues (Samson et al., 1985, 1987b; Weigel et al., 1988; Kovacevic et al., 1989).

This latter achievement has made available convenient sources for preparing large amounts of purified biosynthetic enzymes. This availability will greatly accelerate the understanding of these enzymes and will augment the in-vitro synthesis of novel beta-lactams (see publications by Baldwin, Abraham, and colleagues; Westlake, Jensen, and colleagues; Martin and colleagues).

Recently the genes for acylCoA:IPN acyltransferase from *P. chrysogenum* (Veenstra et al., 1989; Barredo et al., 1989b) and *A.nidulans* (Montenegro et al., 1990; Tobin et al., manuscript in preparation) have been cloned and express in *E. coli*. The gene(s) encoding ACV synthetase in penicillin and cephalosporin-producing fungi is currently the target of cloning experiments, and success can be expected in the near future.

Transformation systems using integrative vectors carrying dominant markers with homologous promoters have allowed the introduction of cloned antibiotic biosynthetic genes into both *A. chrysogenum* and *P. chrysogenum*. The "*pyrG*" transformation system is efficient enough to permit the direct selection of genes by the complementation of blocked mutants in *P. chrysogenum*. In the future, improvements in the existing transformation systems and the development of additional transformation systems for penicillin/cephalosporin-producing fungi will undoubtedly occur.

In the future, other "secondary" and "primary" promoters will be isolated and coupled to a variety of open reading frames. Already an expandase open reading frame has been coupled to a *P. chrysogenum* promoter allowing the expression of ring-expansion activity in *P. chrysogenum* (Cantwell et al., 1990).

The full potential of amplifying gene expression to increase penicillin and cephalosporin production will only be achieved when we manage to understand the regulatory mechanisms (Demain et al., 1983; Martin et al., 1986) that control in vivo expression of the biosynthetic genes involved in producing these antibiotics. Some advance has been made in understanding the molecular mechanisms of carbon catabolite regulation and phosphate control, but a continued research effort is required. Nevertheless, amplication of cloned penicillin and cephalosoporin biosynthetic genes is already being studied in fungi (Martin, unpublished), and practical results using amplified genes have already been observed (Skatrud et al., 1989; Foglesong, 1988).

NOTE ADDED IN PROOF

This review was completed in early 1989. Since then considerable advances have been made in our knowledge of the penicillin gene cluster and in the organization of the cephalosporin and cephamycin biosynthetic genes. The three genes of the penicillin biosynthetic pathway are linked in a single cluster [Diez, B., Barredo, J.L., Alvarez, E., Cantoral, J.M., van Solingen, P., Groenen, M.A.M., Veenstra, A.E., and Martin, J.F. (1989). Two genes involved in penicillin biosynthesis are linked in a 5.1 kb *Sal*I fragment in the genome of *Penicillium chrysogenum. Mol. Gen. Genet. 218*:572-576; Smith, D.J., Brunham, M.K.R., Edwards, J., Earl, A.J., and Turner, G. (1990). Cloning and heterologous expression of the penicillin biosynthetic gene cluster from *Penicillium chrysogenum. Bio/Technology 8*:39-41; Diez, B. et al., submitted for publication]. The *npe*A locus of *A. nidulans* consists of three genes that appear to correspond with the three biosynthetic genes cloned from *P. chrysogenum* DNA [MacCabe, A.P., Riach, M.B.R., Unkless, S.E., and Kinghorn, J.F. (1990). The *Aspergillus npe*A consists of three contigous genes required for penicillin biosynthesis. *EMBO J. 9*:279-287].

REFERENCES

Abraham, E.P., Huddleston, J.A., Jayatilake, G.S., O'Sullivan, J., and White, R. L. (1981). Conversion of δ-(L-α-aminoadipyl)-L-cysteinyl-D-valine to iso-penicillin N in cell-free extracts of *Cephalosporium acremonium*. In *Recent Advances in the Chemistry of Beta-Lactam Antibiotics*, Gregory, G.E., ed., The Royal Society of Chemistry, Burlington House, London, p. 125-134.

Abraham, E.P. (1986). Biosynthesis of penicillins and cephalosporins. In *Beta-Lactam Antibiotics for Clinical Use*, Queener, S.F., Weber, J.A., and Queener, S.W., eds., Marcel Dekker, New York, p. 103-116.

Alvarez, E., Cantoral, J.M., Barredo, J.L., Diez, B., and Martin, J.F. (1987). Purification to homogeneity and characterization of acylcoenzyme A:6-amino-penicillanic acid acyltransferase of *Penicillium chrysogenum. Antimicrob. Agents Chemother.31*:1675-1682.

Baldwin, J.E., White, R.L., John, E.-M.M., and Abraham, E.P. (1982). Stoichiometry of oxygen consumption in the biosynthesis of isopenicillin N from a tripeptide. *Biochem. J. 203*:791-793.

Baldwin, J.E. and Abraham, E. (1988). The biosynthesis of penicillins and cephalosporins. *Natural Prod. Rep. 1988*:129-145.

Baldwin, J.E., Adlington, R.M., Cortes, J.B., Crabbe, J.C., Crouch, N.P., Keeping, J.W., Knight, G.C., Schofield, C.J., Ting, H.H., Vallejo, C.A., Thorniley, M., and Abraham, E.P., (1987a). Purification and initial characterization of an enzyme with deacetoxycephalosporin C synthetase and hydroxylase activities, *Biochem. J. 245*:831-841.

Baldwin, J.E., Killin, S.J., Pratt, A.J., Sutherland, J.D., Turner, N.J., Crabbe, J.C., Abraham, E.P., and Willis, A.C. (1987b). Purification and characterization of cloned isopenicillin N synthetase. *J. Antibiot. 40*:652-659.

Ballance, D.J., Buxton, F.P., and Turner, G. (1983). Transformation of *Aspergillus nidulans* by the orotidine-5′-phosphate decarboxylase gene of *Neurospora crassa*. *Biochem. Biophys. Res. Commun. 112*:284-289.

Ballance, D.J. and Turner, G. (1985). Development of a high-frequency transforming vector for *Aspergillus nidulans*. *Gene 36*:321-331.

Banko, G., Demain, A.L., and Wolfe, S. (1987). δ-(L-α-aminoadipyl-L-cysteinyl-D-valine: a multifunctional enzyme with broad substrate specificity for synthesis of penicillin and cephalosporin precursors. *J. Am. Chem. Soc. 109*:2858-2860.

Barredo, J.L., Alvarez, E., Cantoral, J.M., Diez, B., and Martin, J.F. (1988). A glucokinase deficient mutant of *Penicillium chrysogenum* is derepressed in glucose catabolite regulation of both β-galactosidase and penicillin biosynthesis. *Antimicrob. Agents Chemother. 32*:1061-1067.

Barredo, J.L., Cantoral, J.M., Alvarez, E., Diez, B., and Martin, J.F. (1989a). Cloning, sequence analysis and transcriptional study of the isopenicillin N synthase of *Penicillium chrysogenum* AS-P-78. *Mol. Gen. Genet. 216*:91-98.

Barredo, J.L., van Solingen, P., Diez, B., Alvarez, E., Cantoral, J.M., Kattevilder, A., Smaal, E.B., Groenen, M.A.M., Veenstra, A.E., and Martin, J.F. (1989b). Cloning and characterization of the acyl-coenzyme A:6-aminopenicillanic-acid-acyltransferase gene of *Penicillium chrysogenum*. *Gene 83*:291-300.

Barredo, J.L., Diez, E., Alvarez, E., and Martin, J.F. (1989c). Large amplification of a 35-kb DNA fragment carrying two penicillin biosynthetic genes in high penicillin producing strains of *Penicillium chrysogenum*. *Curr. Genet. 16*:453-459.

Beri, R.K. and Turner, G. (1987). Transformation of *Penicillium chrysogenum* using the *Aspgerillus nidulans amdS* gene as a dominant selective marker. *Curr. Genet. 11*:639-641.

Brewer, S.J., Farthing, J.E., and Turner, M.K. (1977). The oxygenation of the 3-methyl group of 7-beta-5-D-adipamido-3-methylceph-3-em-4-carboxylic acid, deacetoxycephalosporin C by extracts of *Acremonium chrysogenum*. *Biochem. Soc. Trans. 5*:1024-1026.

Brewer, S.J., Taylor, P.M., and Turner, M.K. (1980). An adenosine triphosphate-dependent carbamoylphosphate-3-hydroxmethyl-cephem O-carbamoyltransferase from *Streptomyces*. *Biochem. J. 185*:555-564.

Brunner, V.R., Rohr, M., and Zimmer, M. (1968). Zur bioxynthese des penicillins. *Hoppe-Seyler's Z. Physiol. Chem. 349*:95-103.

Bull, J.H., Smith, D.J., and Turner, G. (1988). Transformation of *Penicillium chrysogenum* with a dominant selectable marker. *Curr. Genet.* *13*:377-382.

Cagloti, M.T. and Sermonti, G. (1956). A study of the genetics of penicillin-producing capacity in *Penicillium chrysogenum*. *J. Gen. Microbiol.* *14*:38-46.

Cantoral, J.M., Diez, B., Barredo, J.L., Alvarez, E., and Martin, J.F. (1987). High frequency transformation of *Penicillium chrysogenum*. *Bio/Technology* *5*:494-497.

Cantwell, C.A., Beckman, R.J., Dotzlaf, J.E., Fisher, D.L., Skatrud, P.L., Yeh, W.K., and Queener, S.W. (1990). Cloning and expression of a hybrid *S. clavuligerus cefE* gene in *P. chrysogenum*. *Curr. Genet.* *17*:213-221.

Carr, L.G., Skatrud, P.L., Scheetz, M.E., Queener, S.W., and Ingolia, T.D., (1986). Cloning and expression of isopenicillin N synthetase gene from *Penicillium chrysogenum*. *Gene 48*:257-266.

Carr, L.G., Skatrud, P.L., Ingolia, T.D., and Queener, S.W. (1987). Organization of the 5.8S, 16-18S, and 23-28S ribosomal RNA genes of *Cephalosporium acremonium*. *Curr. Genet. 12*:209-214.

Carr, L.G., Ingolia, T.D., Kovacevic, S., Miller, J.R., Queener, S.W., Samson, S.M., Skatrud, P.L., Tobin, M.B., and Weigel, J. (1988). Cloning of the β-lactam biosynthetic genes from *Streptomyces* and fungi. In *Abstracts of the 4th ASM Conference on the Genetics and Molecular Biology of Industrial Microorganisms*, Am. Soc. Microbiol., Washington, D.C., *P39*, p. 21.

Castro, J.M., Liras, P., Laiz, L., Cortes, J., and Martin, J.F. (1988). Purification and characterization of the isopenicillin N synthase of *Streptomyces lactamdurans*. *J. Gen. Microbiol. 134*:133-141.

Chapman, J.L., Skatrud, P.L., Ingolia, T.D., Samson, S.M., Kaster, K.R., and Queener, S.W. (1987). Recombinant DNA studies in *Cephalosporium acremonium*. In *Developments in Industrial Microbiology*, Vol. 27, Pierce, G., ed., Soc. Industr. Microbiol., Washington, D.C., p. 165-174.

Chiang, S.-J. (1988). Genetic transformation of *Penicillium chrysogenum* and plasmid chromosomal DNA interaction. In *Abstracts of the ASM Conference on the Genetics and Molecular Biology of Industrial Microorganisms*. Am. Soc. Microbiol., Washington, D. C., abstract IV-11, p. 38.

Cole, D.S., Holt, G., and MacDonald, K.D. (1976). Relationship of the genetic determination of impaired penicillin production in naturally occurring strains to that in induced mutants of *Aspergillus nidulans*. *J. Gen. Microbiol. 96*:423-426.

Cortes, J., Martin, J.F., Castro, J.M., Laiz, L., and Liras, P. (1987). Purifications and characterization of a 2-oxoglutarate-linked ATP-independent deacetoxycephalosporin C synthase of *Streptomyces lactamdurans*. *J. Gen. Micro. 133*:3165-3174.

Demain, A.L. (1983). Biosynthesis of beta-lactam antibiotics. In *Antibiotics Containing the Beta-Lactam Structure*, Vol. 1, Demain, A.L., and Salomon, N.A., eds., Springer-Verlag, Berlin, p. 189.

Demain, A.L., Aharonowitz, Y., and Martin, J.F. (1983). Metabolic control of secondary biosynthetic pathways. In *Biochemistry and Genetic Regulation of Commericially Important Antibiotics*. Vining, L., ed., Addison-Wesley, Reading, MA, p. 49.

Diez, B., Alvarez, E., Cantoral, J.M., Barredo, J.L., and Martin, J.F. (1987). Selection and characterization of *pyr*G mutants of *Penicillium chrysogenum* by resistance to 5'-fluoroorotic acid. *Curr. Genet. 12*:277-282.

Ditchburn, P., Holt, G., and MacDonald, K.D. (1976). The genetic location of mutations increasing penicillin yield in *Aspergillus nidulans*. In *Proceedings of the Second International Symposium on the Genetics of Industrial Microorganisms,* MacDonald, K.D., ed., Academic Press, London, p. 213-227.

Dotzlaf, J.E. and Yeh, W.K. (1987). Copurification and characterization of deacetoxycephalosporin C synthetase/hydroxylase from *Cephalosporium acremonium*. *J. Bacteriol. 169*:1611-1618.

Dotzlaf, J.E. and Yeh, W.K. (1989). Purification and properties of deacetoxycephalosporin C synthase from recombinant *Escherichia coli* and its comparison with the native enzyme purified from *Streptomyces clavuligerus*. *J. Biol. Chem. 264*:10219-10227.

Drew, S.W. and Demain, A.L. (1977). Effect of primary metabolites on secondary metabolism. *Ann. Rev. Microbiol. 31*:343-356.

Edwards, G.F., Holt, G., and MacDonald, K.D. (1974). Mutants of *Aspergillus nidulans* impaired in penicillin biosynthesis. *J. Gen. Microbiol. 84*:420-422.

Fawcett, P.A., Usher, J.J., Huddleston, J.A., Bleany, R.C., Nisbet, J.J., and Abraham, E.P. (1976). Synthesis of δ-(α-aminoadipyl)-cysteinyl-valine and its role in penicillin biosynthesis. *Biochem. J. 157*:651-660.

Foglesong, M.A. (1988). *Federal Register 52*:48660.

Fujisawa, Y. and Kanzaki, T. (1975). Role of acetylcoenzyme deacetylcephalosporin C acetyltransferase in cephalosporin C biosynthesis by *Cephalosporium acremonium*. *Agric. Biol. Chem. 39*:2043-2048.

Fujisawa, Y., Shirafuji, H., Kida, M., Nara, K., Yoneda, M., and Kanzaki, T. (1973). New findings on cephalosporin C biosynthesis. *Nature New Biol. 246*: 153-155.

Fujisawa, Y., Shirafuji, H., and Kanzaki, T. (1975a). Deacetylcephalosporin C formation by cephalosporin C acetylhydrolase induced in a *Cephalosporium acremonium* mutant. *Agric. Biol. Chem. 39*:1303-1305.

Fujisawa, Y., Shirafuji, H., Kida, M., Nara, K., Yoneda, M., and Kanzaki, T. (1975b). Accumulation of deacetylcephalosporin C by cephalosporin C negative mutants of *Cephalosporin acremonium*. *Agric. Biol. Chem. 39*:1295-1302.

Gunsalus, I.C., Pederson, T.C., and Sligar, S.G. (1975). Oxygenase-catalyzed biological hydroxylations. In *Annual Reviews of Biochemistry,* Vol. 44, Snell, E.E., ed., Annual Reviews, Palo Alto, CA, p. 391-393.

Hamlyn, P.F. and Ball, C. (1979). Recombination studies with *Acremonium chrysogenum*. In *Third International Symposium on the Genetics of Industrial Microorganisms,* Sebek, D.K., and Laskin, A.I., eds., Am. Soc. for Microbiol., Washington, D.C., p. 185-191.

Hollander, I.J., Shen, Y.Q., Heim, J., Demain, A.L., and Wolfe, S. (1984). Pure enzyme catalyzing penicillin biosynthesis. *Science 224*:610-612.

Holt, G., Edwards, G.F., and MacDonald, K.D. (1976). The genetics of mutants impaired in the biosynthesis of penicillin. In *Proceedings of the Second International on the Symposium. Genetics of Industrial Microorganisms.* MacDonald, K.D., ed., Academic Press, London, p. 199-211.

Hopwood, D. and Chater, K.F. (1984). Streptomycetes. In *Genetics and Breeding of Industrial Microorganisms*, Ball, C., ed., CRC Press, Boca Raton, FL, p. 7.

Hori, H. and Osawa, S. (1979). Evolutionary changes in 5s RNA secondary structure and a phylogenic tree of 5s RNA species. *Proc. Natl. Acad. Sci. USA 76*: 381-385.

Ingolia, T.D., and Queener, S.W. (1989). Beta-lactam biosynthetic genes. *Med. Res. Rev. 9*:245-263.

Jayatilake, G.S., Huddleston, J.A., and Abraham, E.P. (1981). Conversion of isopenicillin N into penicillin N in cell-free extracts of *Cephalosporium acremonium*. *Biochem. J. 194*:645-648.

Jensen, S.E., Westlake, D.W., and Wolfe, S. (1982). Cyclization of δ-(L-α-aminoadipyl)-L-cysteinyl-D-valine to penicillin by cell-free extracts of *Streptomyces clavuligerus*. *J. Antibiot. 35*:483-490.

Jensen, S.E., Westlake, D.W.S., and Wolfe, S. (1983). Partial purification and characterization of isopenicillin N epimerase activity from *Streptomyces clavuligerus*. *Can. J. Microbiol. 29*:1526-1531.

Jensen, S.E., Westlake, D.W.S., and Wolfe, S. (1985). Deacetoxycephalosporin C synthetase and deacetoxycephalosporin C hydroxylase are two separate enzymes in *Streptomyces clavuligerus*. *J. Antibiot. 38*:263-265.

John, M.A. and Peberdy, J.F. (1984). Transformation of *Aspergillus nidulans* using the *argB* gene. *Enzyme Microb. Technol. 6*:386.

Johnstone, I.L. (1985). Transformation of *Aspergillus nidulans*. *Microbiol. Sci. 2*:307.

Johnstone, I.L., Hughes, S.G., and Clutterbuck, A.J. (1985). Cloning an *Aspergillus nidulans* developmental gene by transformation. *EMBO J. 4*:1307-1311.

Kaster, K.R., Burgett, S.G., and Ingolia, T.D. (1984). Hygromycin B resistance as dominant selectable marker in yeast. *Curr. Genet. 8*:353-358.

Kolar, M., Punt, P.J., van den Hondel, C.A.M.J.J., and Schwab, H. (1988). Transformation of *Penicillium chrysogenum* using dominant selection markers and expression of an *Escherichia coli lacZ* fusion gene. *Gene 62*:127-134.

Komatsu, K., Mizuno, M., and Kodaira, R., (1975). Effect of methionine on cephalosporin C and penicillin production by a mutant of *Acremonium chrysogenum*. *J. Antibiot. 28*:881-888.

Kovacevic, S., Weigel, B.J., Tobin, M.B., Ingolia, T.D., and Miller, J.R. (1989). Cloning, characterization, and expression in *Escherichia coli* of the *Streptomyces clavuligerus* gene encoding deacetoxycephalosporin C synthetase. *J. Bacteriol. 171*:754-760.

Kupka, J., Shen, Y.Q., Wolfe, S., and Demain, A.L. (1983). Studies on the ring cyclization and ring expansion enzymes of β-lactam biosynthesis in *Acremonium chrysogenum*. *Can. J. Microbiol. 29*:488-496.

Laiz, L., Liras, P., Castro, J.M., and Martin, J.F. (1990). Purification and characterization of the isopenicillin N epimerase from *Nocardia lactamdurans*. *J. Gen. Microbiol. 136*:663-671.

Lemke, P. and Nash, C. (1971). Mutations that affect antibiotic synthesis by *Cephalosporium acremonium*. *Can. J. Microbiol. 18*:255-259.

Leskiw, B.K., Aharonwitz, Y., Mevarech, M., Wolfe, S., Vining, L.C., Westlake, D.W.S., and Jensen, S.E. (1988). Cloning and nucleotide sequence determination of the isopenicillin N synthetase gene from *Streptomyces clavuligerus*. *Gene 62*:187-196.

Li, W.-H., Luo, C.-C., and Wu, C.-I. (1985). Evolution of DNA sequences. In *Molecular Evolutionary Genetics*, MacIntyre, R.J., ed., Plenum Press, New York, p. 1-94.

Liersch, M., Nuesch, J., and Treichler, H.J. (1976). Final steps in the biosynthesis of cephalosporin C. In *Second International Symposium on the Genetics of Industrial Microorganisms*. MacDonald, K.D., ed., Academic Press, London, p. 179-195.

Lubbe, C., Wolfe, S., and Demain, A.L. (1986). Isopenicillin N epimerase activity in a high cephalosporin-producing strain of *Cephalosporium acremonium*. *Appl. Microbiol. Biotechnol. 23*:367-368.

MacDonald, K.D. (1983). Fungal genetics and antibiotic production. In *Biochemistry and Genetic Regulation of Commercially Important Antibiotics*, L.C. Vining, ed., Addison-Wesley, Reading, MA, p. 25.

Mahro, B. and Demain, A.L. (1987). In vivo conversion of penicillin N into cephalosporin type antibiotic by a non-producing mutant of *Streptomyces clavuligerus*. *Appl. Microbiol. Biotechnol. 27*:272-275.

Makins, J.F., Holt, G., and MacDonald, K.D. (1980). Co-synthesis of penicillin following treatment of mutants of *Aspergillus nidulans* impaired in antibiotic production with lytic enzymes. *J. Gen. Microbiol. 119*:397-404.

Makins, J.F., Allsop, A., and Holt, G. (1981). Intergeneric cosynthesis of penicillin by strains of *Penicillin chrysogenum, P. notatum* and *Aspergillus nidulans, J. Gen. Microbiol. 122*:339-343.

Makins, J.F., Holt, G., and MacDonald, K.D. (1983). The genetic location of three mutations impairing penicillin production in *Aspergillus nidulans*. *J. Gen. Microbiol. 129*:3027-3033.

Martin, J.F. and Demain, A.L. (1980). Control of antibiotic biosynthesis. *Microbiol. Rev. 44*:232-251.

Martin, J.F. and Liras, P. (1981). Biosynthetic pathways of secondary metabolites in industrial microorganisms. In *Biotechnology: A Comprehensive Treatise*, Vol. 1. *Microbial Fundamentals,* Rehm, H.J., and Reed, G., eds., Verlag Chemie, Weinheim, p. 211.

Martin, J.F. and Gil, J.A. (1982). Cloning and expression of antibiotic production genes. *Bio/Technology 2*:63-72.

Martin, J.F. and Aharonowitz, Y. (1983). Regulation of biosynthesis of β-lactam antibiotics. In *Antibiotics Containing the β-Lactam Structure*, Vol. 1, Demain, A.L., and Solomn, N.A., eds., Springer-Verlag, Berlin, p. 229-254.

Martin, J.F. and Liras, P. (1985). Biosynthesis of β-lactam antibiotics: design and construction of overproducing strains, *Trends Biotechnology 3*:39-44.

Martin, J.F., Lopez-Nieto, J.M., Castro, J.M., Cortes, J., Romero, J., Ramos, F.R., Cantoral, J.M., Alvarez, E., Dominguez, M.G., Barredo, J.L., and Liras, P. (1986). Enzymes involved in beta-lactam biosynthesis controlled by carbon and nitrogen regulation. In *Regulation of Secondary Metabolite Formation,* Kleinkauf, H., von Dohren, H., Dronauer, H., and Nesseman, G., eds., VCH Verlag, Weinheim, p. 41-75.

Martin, J.F., Diez, B., Alvarez, E., Barredo, J.L., and Cantoral, J.M. (1987). Development of a transformation system in *Penicillium chrysogenum*: Cloning of genes involved in penicillin biosynthesis. In *Genetics of Industrial Microorganisms*, Alacevic, M., Hranueli, D., and Toman, Z., eds., Pliva, Zagreb, p. 297-308.

Martin, J.F. and Liras, P. (1989). Enzymes involved in penicillin, cephalosporin and cephamycin biosynthesis. In *Advances in Biochemical Engineering*. Fiechter, A., ed., Springer Verlag, Berlin, pp. 153-187.

Meesschaert, B., Adriens, P., and Eyssen, H. (1980). Studies on the biosynthesis of isopenicillin N with cell-free preparation of *Penicillium chrysogenum*. *J. Antibiot.* 33:722-730.

Miller, J.R. and Ingolia, T.D. 1989. Cloning and characterization of beta-lactam biosynthetic genes. *Mol. Microbiol.* 3:689-695.

Montenegro, E., Barredo, J.L., Gutierrez, S., Diez, B., Alvarez, E., and Martin, J.F. (1990). Cloning, characterization of the acyl-CoA:6-amino penicillanic acid acyltransferase gene of *Aspergillus nidulans* and linkage to the isopenicillin N synthase gene. *Mol. Gen. Genet.*, in press.

Nash, C.H., de la Higuera, N., Neuss, N., and Lemke, P.A. (1974). Application of biochemical genetics to the biosynthesis of β-lactam antibiotics. *Develop. Indust. Microbiol.* 15:114-123.

Neuss, N., Nash, C.H., Lemke, P.A., and Grutzner, J.B. (1971). The use of carbon-13 nuclear magnetic resonance (Cmr) spectroscopy in biosynthetic studies. Incorporation of carboxyl and methyl carbon-13 labeled acetates into cephalosporin C. *J. Am. Chem. Soc.* 93:2337-2339.

NCIUB, 1978. Isomerases. In *Recommendations (1978) of the Nomenclature Committee of the International Union of Biochemistry*. Academic Press, New York, p. 412-413.

Normansell, P.J.M., Normansell, I.D., and Holt, G. (1979). Genetic and biochemical studies of mutants of *Penicillium chrysogenum* impaired in penicillin production. *J. Gen. Microbiol.* 112:113-126.

Nuesch, J., Treichier, H.J., and Liersch, M. (1973). The biosynthesis of cephalosporin C. In *The Genetics of Industrial Microorganisms. First International Symposium,* Vanek, Z., Hostalek, Z., and Cudlin, J., eds., Academia, Prague, p. 309-334.

O'Sullivan, J. and Abraham, E.P. (1980). The conversion of cephalosporins to 7-α-methoxycephalosporins by cell-free extracts of *Streptomyces clavuligerus*. *Biochem. J.* 186:613-616.

Pang, C.P., Chakravarti, B., Adlington, R.M., Ting, H.H., White, R.L., Jayatilake, G.S., Baldwin, J.E., and Abraham, E.P. (1984). Purification of isopenicillin N synthase. *Biochem. J.* 222:789-795.

Perez-Martinez, G. and Peberdy, J.F. (1985). Production of cephalosporin C and its intermediates by raised-titre strains of *Acremonium chrysogenum*. *Enzyme Microb. Technol.* 7:389.

Penalva, M.A., Tourino, A., Pantino, C., Sanchez, F., Fernandez-Sousa, J.M., and Rubio, V. (1985). In *Molecular Genetics of Filamentous Fungi*, Timberlake, W.E., ed., Alan R. Liss, New York, p. 59-68.

Picknett, T.M., Saunders, G., Ford, P., and Holt, G. (1987). Development of gene transfer system for *Penicillium chrysogenum. Curr. Genet. 12*:449-455.

Pruess, D.L. and Johnson, M.J. (1967). Penicillin acyltransferase in *Penicillium chrysogenum. J. Bacteriol. 94*:1502-1508.

Queener, S.W. (1989). Molecular biology of penicillin and cephalosporin B synthesis. *Antimicrob. Agents Chemother. 34*:943-948.

Queener, S.W., Capone, J.J., Radue, A.B., and Nagarajan, R. (1974). Synthesis of deacetoxycephalosporin C by a mutant of *Cephalosporium acremonium. Antimicrob. Agents. Chemother. 6*:334-337.

Queener, S.W. and Neuss, N. (1982). The biosynthesis of beta-lactam antibiotics. In *Chemistry and Biology of Beta-Lactam Antibiotics.* Vol. 3, Morin, R.B., and Gorman, M., eds., Academic Press, New York, p. 1-82.

Queener, S.W., Wilkerson, S., Tunin, D.R., McDermott, J.R., Chapman, J.L., Nash, C., Platt, C., and Westpheling, J. (1984). Cephalosporin C fermentation: biochemical and regulatory aspects of sulfur metabolism. In *Biotechnology of Industrial Antibiotics*, Vandamme, E., ed., Marcel Dekker, New York, p. 141-170.

Queener, S.W., Ingolia, T.D., Skatrud, P.L., Chapman, J.L., and Kaster, K.R. (1985). A system for genetic transformation of *Cephalosporium acremonium.* In *Microbiology—1985*, Leive, L., ed., Am. Soc. Microbiol., Washington, D.C., p. 468-472.

Queener, S.W., Skatrud, P.L., Ingolia, T.D., Yeh, W.K., Tietz, A., and McGilvray, D. (1988). Recombinant studies in *Cephalosporium acremonium* and *Penicillium chrysogenum.* In *Abstracts of the Fourth ASM Conference on the Genetics and Molecular Biology of Industrial Microorganisms, Am. Soc. Microbiol.,* Washington, D.C., *P40*, p. 22.

Ramon, D., Carramolino, L., Patino, C., Sanchez, F., and Penalva, M.A. (1987). Cloning and characterization of the isopenicillin N synthetase gene mediating the formation of the β-lactam ring in *Aspergilllus nidulans. Gene 57*:171-181.

Ramos, F.R., Lopez-Nieto, M.J., and Martin, J.F. (1985). Isopenicillin N synthase of *Penicillin chrysogenum*, an enzyme that converts δ-(L-α-aminoadipyl)-L-cysteinyl)-D-valine to isopenicillin N. *Antimicrob. Agents Chemother. 27*:380-387.

Ramos, F., Lopez-Nieto, M.J., and Martin, J.F. (1986). Coordinate increase of isopenicillin N synthetase, isopenicillin N epimerase and deacetoxycephalosporin C synthetase in a high cephalosporin-producing mutant of *Acremonium chrysogenum. FEMS Microbiol. Lett. 35*:123-127.

Ramsden, M., McQuade, B.A., Saunders, K., Turner, M.K., and Hafford, S. (1989). Characterization of a loss-of-function mutation in the isopenicillin N synthase gene of *Acremonium chrysogenum. Gene 85*:267-273.

Revilla, G., Ramos, F.R., Lopez-Nieto, M.J., Alverez, E., and Martin, J.F., (1986). Glucose represses formation of δ-(L-a-aminoadipyl)-L-cysteinyl-D-valine and isopenicillin N synthase but not penicillin acyltransferase in *Penicillium chrysogenum. J. Bacteriol. 168*:947-952.

Samson, S.M., Belagaje, R., Blankenship, D.T., Chapman, J.L., Perry, D., Skatrud, P.L., Vanfrank, R.M., Abraham, E.P., Baldwin, J.E., Queener, S.W., and Ingolia, T.D. (1985). Isolation, sequence determination and expression in *Escherichia coli* of the isopenicillin N synthetase gene from *Cephalosporium acremonium. Nature 318*:191-194.

Samson, S.M., Chapman, J.L., Belagaje, R., Queener, S.W., and Ingolia, T.D. (1987a). Analysis of the role of cysteine residues in isopenicillin N synthetase activity by site-directed mutagenesis. *Proc. Natl. Acad. Sci. USA 84*:5705-5709.

Samson, S.M., Dotzlaf, J.E., Slisz, M.L., Becker, G.W., Van Frank, R.M., Veal, L.E., Yeh, W.K., Miller, J.R., Queener, S.W., and Ingolia, T.D. (1987b). Cloning and expression of the fungal expandase/hydroxylase gene involved in cephalosporin biosynthesis, *Bio/Technology 5*:1207-1214.

Sanchez, F., Lozano, M., Rubio, V., and Penalva, M.A. (1987). Transformation of *Penicillium chrysogenum. Gene 51*:97-102.

Scheidegger, A., Kuenzi, M.T., and Nuesch, J. (1984). Partial purification and catalytic properties of a bifunctional enzyme in the biosynthetic pathway of β-lactams in *Acremonium chrysogenum. J. Antibiot. 37*:522-531.

Sermonti, G. (1956). Complementary genes which affect penicillin yields. *J. Gen. Microbiol. 15*:599-608.

Sermonti, G. (1959). Genetics of penicillin production. *Ann. NY Acad. Sci. 81*: 950-966.

Shen, Y., Wolfe, S., and Demain, A.L. (1986). Levels of isopenicillin N synthetase and deacetoxycephalosporin C synthetase in *Cephalosporium acremonium* strains producing high and low levels of cephalosporin C. *Bio/Technology 4*:61-62.

Shirafuji, H., Fujisawa, Y., Kida, M., Kanzaki, T., and Yoneda, M. (1979). Accumulation of tripeptide derivatives by mutants of *Cephalosporium acremonium. Agric. Biol. Chem. 43*:155-160.

Shirafuji, H. and Yoneda, M. (1982). Accumulation of δ-(L-α-aminoadipyl)-L-cysteinyl-D-valine derivatives by mutants of *Cephalosporium acremonium*. In *Peptide Antibiotics: Biosynthesis and Functions*, Walter de Gruyter, Berlin, p. 97-139.

Simpson, I.N. and Caten, C.E. (1980). Genetics of penicillin titre in lines of *Aspergillus nidulans* selected through recurrent mutagenesis. *J. Gen. Microbiol. 121*:5-16.

Simpson, I.N. and Caten, C.E. (1981). Selection for increased penicillin titre following hybridization of divergent lines of *Aspergillus nidulans. J. Gen. Microbiol. 126*:311-319.

Skatrud, P.L. and Queener, S.W. (1984). Cloning of a DNA fragment from *Acremonium chrysogenum* which functions as an autonomous replication sequence in yeasts. *Curr. Genet. 8*:155-164.

Skatrud, P.L., Queener, S.W., Carr, L.G., and Fisher, D.L. (1987a). Efficient integrative transformation of *Cephalosporium acremonium. Curr. Genet. 12*: 337-348.

Skatrud, P.L., Fisher, D.L., Ingolia, T.D., and Queener, S.W. (1987b). Improved transformation of *Cephalosporium acremonium*. In: *Genetics of Industrial Microorganisms*, Alacevic, M., Hranueli, D., Toman, Z., eds., Publ. Pliva, Zagreb, Yugoslavia, p. 111-119.

Skatrud, P.L., Queener, S.W., Fisher, D.L., and Chapman, J.L. (1987c). Strain improvement studies in *Penicillium chrysogenum* using the cloned *P. chrysogenum* isopenicillin N synthetase gene and the *amdS* gene of *Aspergillus nidulans. Abstracts of the Annual Meeting of the Society of Industrial Microbiology, SIM News 37*:77.

Skatrud, P.L. and Queener, S.W. (1989). An electrophoretic molecular karyotype for an industrial strain of *Cephalosporium acremonium*. *78*:331-338.

Skatrud, P.L., Tietz, A.J., Ingolia, T.D., Cantwell, C.A., Fisher, D.L., Chapman, J.L., and S.W. Queener. (1989). Use of recombinant DNA to improve production of cephalosporin C by *Cephalosporium acremonium*. *Bio/Technology 7*:477-485.

Stahl, U., Leitner, E., and Esser, K. (1987). Transformation of *Penicillium chrysogenum* by a vector containing a mitochondrial origin of replication. *Appl. Microbiol. Biotechnol. 26*:237-241.

Stevens, C.M., Abraham, E.P., Huang, F.C., and Cih, C.J. (1975). (abstract). *FASEB J. 34*:625.

Smith, D.J., Bull, J.H., Edwards, J., and G. Turner. (1988). Amplification of the isopenicillin N synthetase gene in a strain of *Penicillium chrysogenum* producing high levels of penicillin. In *Abstracts of the ASM Conference on the Genetics and Molecular Biology of Industrial Microorganisms*, Am. Soc. Microbiol., Washington, D.C., abstract IV19, p. 40.

Tilburn, J., Scazzochio, C., Taylor, G.G., Zabicky-Zissman, J.H., Lockington, R.A., and Davies, R.W. (1984). Transformation by integration in *Aspergillus nidulans. Gene 26*:205-211.

Turner, M.K., Farthing, J.E., and Brewer, S.J. (1978). The oxygenation of (3-methyl-3H)deacetoxycephalosporin C to (7β-(5-D-aminoadipamido)-3-methyl-ceph-3-em-4-carboxylic acid) to (3-hydroxymethyl-3H) desacetylcephalosporin C by 2-oxoglutarate-linked dioxygenases from *Acremonium chrysogenum* and *Streptomyces clavuligerus. Biochem. J. 173*:839-850.

Tudzynski, P. and Esser, K. (1982). Extrachromosomal genetics of *Acremonium chrysogenum*: II. Development of a mitochondrial DNA hybrid vector replicating in *Saccharomyces cerevisiae. Curr. Genet. 6*:153-164.

Usui, S., and Yu, C.-A. (1989). Purification and properties of isopenicillin N epimerase. *Biochem. Biophys. Acta*, in press.

van Liempt, H., von Dohren, H., and Kleinkauf, H. (1989). δ-(L-α-aminoadipyl)-L-cysteinyl-D-valine synthetase from *Aspergillus nidulans. J. Biol. Chem. 264*:3680-3684.

Veenstra, A. E., Van Solingen, P. V., Huininga-Muurling, H., Koekman, B.P., Groenen, M.A.M., Smaal, E.B., Kattevilder, A., Alvarez, E., Barredo, J.L., and Martin, J.F. (1989). Cloning of penicillin biosynthesis genes. In *Abstracts of the ASM Conference on the Genetics and Molecular Biology of Industrial Microorganisms*, Am. Soc. Microbiol., Washington, D.C., p. 262-269.

Wernars, K., Goosen, T., Wennekes, L.M.J., Visser, J., Bos, C.J., Van Del Hondel, C.A.M.J.J., and Pouwels, P.H. (1985). Gene amplification in *Aspergillus nidulans* by transformation with vectors containing the *amdS* gene. *Curr. Genet. 9*:361-368.

Weigel, B.J., Burgett, S.G., Chen, V.J., Skatrud, P.L., Frolik, C.A., Queener, S.W., and Ingolia, T.D. (1988). Cloning and expression E. coli of isopenicillin N synthetase genes from *Steptomyces lipmanii* and *Aspergillus nidulans. J. Bacteriol. 170*:3817-3826.

Whiteman, P.A., Abraham, E.P., Baldwin, J.E., Fleming, M.D., Schofield, C.J., Sutherland, J.D., and Willis, A.C. (1990). Acyloenzyme A:G-aminopenicillanic acid acyltransferase from *Penicillium chrysogenum* and *Aspergillus nidulans. FEBS Lett. 262*:342-344.

Wyngaarden, J.B. 1988. *Federal Register 53*:43411.

Yelton, M.M., Timberlake, W.E., and van den Hondel, C.A.M.J.J. (1985). A cosmid for selecting genes by complementation in *Aspergillus nidulans*: selection of the developmentally regulated yA locus. *Proc. Natl. Acad. Sci. USA 82*: 834-838.

Yoshida, M., Konomi, T., Kohsaka, M., Baldwin, J.E., Herlhen, S., Singh, P., Hunt, N.A., and Demain, A.L. (1978). Cell-free ring expansion of penicillin N to deacetoxycephalosporin C by *Acremonium chrysogenum* CW-19 and his mutants. *Proc. Natl. Acad. Sci. USA 75*:6253-6257.

8

Molecular Biology of Lignin Peroxidases from *Phanerochaete chrysosporium*

Erika L. F. Holzbaur,* Andrawis Andrawis,† and Ming Tien
The Pennsylvania State University, University Park, Pennsylvania

I. INTRODUCTION

Knowledge on lignin biodegradation has grown with the discovery of lignin-degrading enzymes in 1981 (Tien and Kirk, 1983; Glenn et al., 1983). This finding has allowed researchers to apply modern methods of biochemistry, biophysics, and molecular biology to the study of lignin biodegradation, research that has lead to the recent cloning and expression of the lignin peroxidase genes. This review will focus on the molecular biology of lignin-degrading enzymes. The progress of the last 2 years will be summarized, as well as the most recent unpublished results from our laboratory. Unpublished findings from other laboratories will be included where appropriate.

The synthesis and degradation of lignin plays a major role in carbon recycling on earth. The biosynthesis of lignin accounts for approximately 25% of the energy that plants capture from the sun. The function of lignin in woody tissue is twofold: It imparts the plant with structural rigidity,

Current affiliations:
*Worcester Foundation for Experimental Biology, Shrewsbury, Massachusetts
†Botany-Life Science Institute, Hebrew University, Jerusalem, Israel

and it protects the cellulose and hemicellulose from microbial attack. Organisms capable of degrading lignin thus possess the selective advantage of gaining access to the cellulose of woody plants. Certain filamentous fungi have evolved an oxidative system that is capable of degrading lignin to the level of CO_2. It is widely accepted that the filamentous fungi represent the major pathway for the biodegradation of lignin in nature.

The chemical nature of the lignin polymer makes it especially resistant to enzymatic attack. It is a highly condensed insoluble polymer of phenylpropane subunits (Goring, 1977). The polymer is synthesized through free radical condensation reactions, which yield a polymer with a variety of carbon-carbon and carbon-oxygen bonds (Adler, 1977). There is no readily apparent repeating linkage in the structure such as is found in all other biological polymers. Thus there is no single bond motif that an enzyme could hydrolyze to depolymerize lignin. From this observation, and without any *a priori* knowledge of lignin-degrading enzymes, one could conclude that lignin-degrading enzymes must be nonspecific in nature.

The biodegradation of lignin has been most extensively studied with the white-rot fungi, and in particular with the filamentous fungus *Phanerochaete chrysosporium*. This fungus is capable of degrading lignin to CO_2. During vegatative growth, the ligninolytic system of this organism is not expressed. The lignin-degrading system is only expressed in response to deprivation of the nutrients nitrogen, carbon, or sulfur (Keyser et al., 1978; Jeffries et al., 1981). The physiological significance of this regulatory system is most likely due to the structure of woody tissue in which cellulose fibrils are embedded in a lignin-hemicellulose matrix (Kirk, 1983). Perhaps once the fungus depletes a layer of cellulose in the woody tissue, the ligninolytic system is induced in order to gain access to the next layer of the more energy-rich cellulose.

The primary component of the ligninolytic system that is produced by *P. chrysosporium* during secondary metabolism is ligninase, or lignin peroxidase. This enzyme has been the subject of detailed characterization. The fungus also secretes Mn-dependent peroxidases in response to nutrient limitation. The lignin peroxidases and the Mn-dependent peroxidases share less than 50% homology, as determined by sequencing of cDNA clones (Tien and Tu, 1987; Gold, 1988). In this review we will be focusing on the lignin peroxidases.

II. LIGNIN PEROXIDASES

A. Isolation of Lignin Peroxidase Isozymes

The lignin-degrading enzymes are a family of isozymes. During nitrogen starvation, lignin peroxidase activity appears by day 3 or 4 of growth and the activity is maximal by days 5-7. The level of enzyme production by the fungus is dependent on culture conditions such as media composition and the degree of agitation. Activity is typically measured by the rate of H_2O_2-dependent oxidation of 3,4-dimethoxybenzyl (veratryl) alcohol to the corresponding aldehyde. Culture preparations with activities ranging from 50 to 1500 U/l have been reported.

The fact that the lignin peroxidase isozymes are secreted makes them relatively simple to purify. The enzyme can be purified in three steps: (a) removal of the mycelium from the extracellular fluid via centrifugation or filtration; (b) concentration of the extracellular fluid; and (c) separation and purification of the isozymes by anion-exchange chromatography. Although purification on conventional DEAE resins yield pure isozymes (Tien and Kirk, 1984), the resolution provided by commercially packed HPLC ion-exchange resin is far superior (Kirk et al., 1986a). A typical elution profile obtained with a Pharmacia Mono Q column is shown in Figure 1.

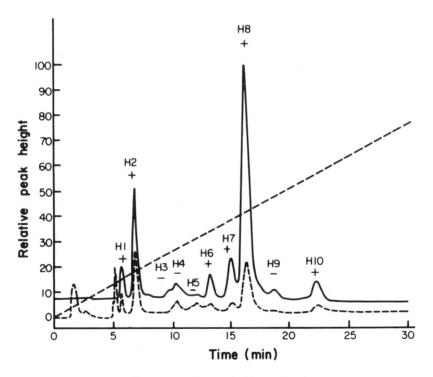

Figure 1 Elution profile from a Pharmacia Mono Q column.

All of the peaks that absorb at 409 nm are hemeproteins. Peaks exhibiting veratryl alcohol-oxidizing activity are marked (+) in Figure 1. The pI values have been determined for the isozymes (Farrell et al., 1989): H1 (pI = 4.6), H2 (pI = 4.4), H6 (pI = 3.7), H7 (pI = 3.6), H8 (pI = 3.5), and H10 (pI = 3.3). Peaks designated H3 (pI = 4.9), H4 (pI = 4.5), and H5 in the elution profile do not have veratryl alcohol-oxidizing activity. These are the manganese-dependent peroxidases, which are catalytically distinct from the lignin peroxidases (Glenn and Gold, 1985).

B. Physical Properties of the Lignin Peroxidase Isozymes

The lignin peroxidase isozyme that has been most extensively characterized is identified as H8, based on ion-exchange elution profiles (Figure 1). This isozyme has a molecular weight of 42,000 (Kirk et al., 1986a). It is a glycoprotein, 15% carbohydrate by weight, and contains one heme iron per enzyme molecule (Kirk et al., 1986a). As a family, the lignin peroxidase isozymes are structurally related; they are all glycoproteins and hemeproteins. Their molecular weights range from 38,000 to 42,000 (Farrell et al., 1989). Structural homology is also demonstrated by the cross-reactivity of polyclonal antibodies (Kirk et al., 1986a) and by peptide mapping (Farrell et al., 1989). Polyclonal antibodies raised to H8 cross-react with the other lignin peroxidase isozymes. However, there is little, if any, reactivity with the manganese-dependent peroxidase (Leisola et al., 1987). Peptide mapping studies using V8 protease showed similar patterns of digestion for the lignin peroxidase isozymes (Farrell et al., 1989). In contrast to an earlier hypothesis (Leisola et al., 1987), the structural similarities are not due to posttranslational modification. As described below, different structural genes have now been isolated for the lignin peroxidase isozymes.

C. Catalytic Properties

The lignin peroxidase isozymes catalyze a wide range of H_2O_2-dependent oxidations in aromatic substrates. Based on kinetic and spectroscopic characterization, ligninase can be catagorized as a peroxidase. Steady-state studies have shown that veratryl alcohol oxidation occurs by a ping-pong mechanism. During the catalytic cycle, the classical peroxidase compound I and compound II enzyme intermediates are formed.

D. Lignin Peroxidase Multiplicity

The multiplicity of the lignin peroxidase isozymes has been noted since the discovery of the enzyme (Tien and Kirk, 1983; Glenn et al., 1983).

The basis for this multiplicity was not clear. Despite small catalytic differences (Farrell et al., 1989), there is as yet no differentiation in roles for the different isozymes. Similar rates of catalysis have been observed for all substrates examined. It was once thought that the various isozymes were partial degradation products of a parental lignin peroxidase. This would be consistent with protein recycling during nitrogen starvation. The first indication that the isozymes are encoded by different RNAs came from amino acid sequence determination at the amino termini of the purified isozymes. The amino-termini sequences of the isozymes are described in Table 1. The isozymes are similar in sequence, but the data clearly indicate differences among the proteins. Analysis of the immunoprecipitation of products from in-vitro translation of poly A RNA isolated from ligninolytic cultures showed multiple polypeptides were formed (Tien and Tu, 1987). Finally, the cloning of lignin peroxidases genes has demonstrated the existence of multiple genes (Tien and Tu, 1987).

III. cDNA CLONES

A. Sequences of the Lignin Peroxidase cDNA Clones

Numerous laboratories have isolated and characterized lignin peroxidases cDNA clones. These include our laboratory, Farrell and co-workers at Repligen/Sandoz, and Reddy and co-workers at Michigan State University. Reddy and co-workers initially published on the isolation of cDNA clones that hybridized to oligonucleotide probes (Zhang et al., 1986). The first report describing the cloning and nucleotide sequencing of a lignin peroxidase cDNA was by Tien and Tu (1987). The cDNA clone from *P. chryso-*

Table 1 Comparison Between the N-terminal Sequences of Lignin Peroxidase Isozymes

---

N-terminal sequence deduced from lignin peroxidase clones

ML-1 (8)   A T C S N G K T V G D A S C C A W F D V L D D I Q Q N L F H
ML-5 (17) A T C S N G A T V S D A S S C A W F D V L D D I Q Q N L F Q
LG-4 (16) V A C P D G V H T A S N A A C C A W F P V L D D I Q Q N L F
LG-5   A T C S N G K V V P A A S C C T W F N V L S D I Q E N L F N

N-terminal sequence of the lignin peroxidase isozymes

H2 (16)   V A C P D G V H T A S N
H6 (17)   A T X A N G A T V G D A S X X A X F D V L D D I Q X N F Q G G Q
H8 (8)   A T W S N G K T V G D A S – E / Q A W F D V L D D
H10 (16)   A T C S N G K V V P

---

sporium (λML-1) was identified as isozyme H8 based on antibody cross-reactivity, and by identity with the amino-terminus sequence of the protein. The 1.4-kb cDNA insert was predicted to encode a mature protein of 37,000 after cleavage of a 28 amino-acid signal peptide. The sequence of two other lignin peroxidase cDNA clones was subsequently published by de Boer et al. (1987). These workers initially identified their clones as encoding isozymes H2 and H8, but subsequently designated them as H2 and H10, which is consistent with the results of our laboratory.

We have isolated and sequenced two additional lignin peroxidase cDNA clones, λML-4 and λML-5 (A. Andrawis, E. Pease, I. Kuan, E.L.F. Holzbaur, and M. Tiem, unpublished observations). The sequences of λML-1 and λML-4 share 98% homology; most of the differences are in the 5' and 3' noncoding regions. The small difference in the coding region results in a change of only one amino acid; Ser (− 10) of λML-4 is replaced by a Leu in λML-1. No differences were observed in the coding region of the mature apoproteins; both λML-1 and λML-4 encode isozyme H8.

The sequence of the λML-5 cDNA was also analyzed. The cDNA encodes a protein of 344 amino acids, 28 of which form the leader sequence. The protein encoded by λML-5 is highly homologous to lignin peroxidase isozyme H8. The deduced amino acid sequence of λML-5 does not match with the experimentally determined amino-termini sequences of H2, H7, or H10. It does match very well with the sequence of H6; the amino terminus sequence is ATX<u>A</u>NGATV<u>G</u>DASXXAXFCVLDDIZXNGQGGQ, in which the amino acids underlined represent mismatches with the DNA sequence. With 25 of the 27 amino acids matching, we suggested that λML-5 endodes isozyme H6. However, the match is not sufficient for an unequivocal assignment.

All of the lignin peroxidase cDNA clones isolated so far are highly homologous. The highest degree of homology is found among λML-1, λML-4, λML-5, and LG-5, which encode the isozymes H8, H8, H6 (putative), and H10, respectively. The least similar of the clones is LG-4, which encodes isozyme H2 (de Boer et al., 1987). The degree of homology is depicted graphically in Figure 2. It is interesting to note that H2 has a higher specific activity (as measured by veratryl alcohol oxidation) than H6, H8, or H10 (Farrell et al., 1989). The plots clearly show that a striking similarity exists among the coding regions of the various lignin peroxidase genes. Despite the similarities in the peroxidase activity and catalytic mechanism, however, there is very little homology between the sequences of the ligninases and other peroxidases.

All of the genes encode a leader sequence of either 27 or 28 predominantly hydrophobic amino acids, which are thought to mediate the secretion

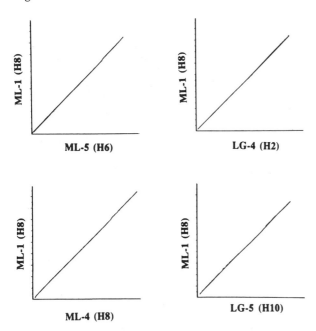

Figure 2 Homology matrix analysis of the various cDNA clones of lignin peroxidase.

of the proteins (Kaput et al., 1983). The leader sequences end in the dibasic residues Lys-Arg. Dibasic residues have been shown to be the site of endoproteolytic scission of precursor peptides such as yeast α-factor (Julius et al., 1984) and peptide hormones (Bell et al., 1983). λML-1, λML-4, λML-5, and LG-5 all contain one potential N-glycosylation site, which follows the general rule of Asn-X-Thr/Ser (Neuberger et al., 1972). LG-4, however, contains two such sites. This finding is consistent with the observation that endo-H treatment of the lignin peroxidase isozymes increases their electrophorectic mobility on SDS-PAGE (Farrell et al., 1989). In addition to potential sites for N glycosylation, there are a multitude of Ser and Thr residues in all of the proteins, which are potential sites for O glycosylation. Our unpublished results provide indirect evidence that the lignin peroxidases are O glycosylated. Treatment of isozyme H8 with a chemical agent specifically used for O deglycosylation (Edge et al., 1981) increased the electrophoretic mobility of the protein on SDS-PAGE.

The eukaryotic consensus signal sequence for polyadenylation, AATAAA (Proudfoot and Brownbee, 1976), was not found in any of the lignin peroxidase clones. The sequence AAATAT, which may be a corresponding sequence in *P. chrysosporium*, is found 12 bp upstream of the poly A tail of λML-1. λML-4 contains the sequence AATCTT 15 bp upstream of the poly A tail. No sequence resembling AATAAA is found in λML-5. The cDNA clone LG-4 contains the sequence AATATA 14 bp upstream of the poly A tail and LG-5 contains the sequence AATACA (de Boer et al., 1987). de Boer et al. (1987) have suggested that AATA(C/T)A may be the consensus polyadenylation signal in *P. chrysosporium*. This sequence, however, is not found in the other lignin peroxidase clones. Therefore, sequence data reveal no clear consensus sequence for polyadenylation.

B. Lignin Peroxidase Structure

Based on results obtained from kinetic (Tien et al., 1986) and physical studies (Kuila et al., 1986), ligninase can be categorized as a peroxidase. Consistent with this classification, two His residues were identified in ligninase that are conserved among the peroxidases and that are referred to as the proximal and distal histidines in reference to the heme group. These residues are thought to be essential for peroxidase activity, based on X-ray crystallography and kinetic analyses. The proximal His is the axial ligand of the heme, a ligand that contributes to the heme's unique catalytic properties (Poulos and Kraut, 1980). The distal His is located near the heme (in the distal pocket) and is thought to be involved in charge stabilization during peroxide cleavage. Peroxide cleavage results in oxidation of the heme by two electrons to yield the enzyme intermediate referred to as compound I (Poulos and Kraut, 1980). Also located in the distal pocket is an Arg, four residues removed from the distal His. This residue is also thought to participate in charge stabilization during peroxide cleavage. The distal and proximal histidine residues of lignin peroxidase are situated at approximately the same location in the amino acid sequence as is found with other peroxidases (Tien and Tu, 1987). This would suggest that the tertiary structure of lignin peroxidase is similar to the structures of other peroxidases, despite limited sequence homology.

A comparison of the predicted amino acid sequence of lignin peroxidase and three other peroxidases, turnip, cytochrome c, and horseradish, indicates that the proximal His (His 175), the distal His (His47), and the Arg (Arg43) residues are conserved (Figure 3). There is very little homology in the residues flanking the proximal His; a feature that may confer upon lignin peroxidase its unique ability to oxidize substrates of high reduction potential (Umezawa and Higuchi, 1985; Kawai et al., 1985; Higuchi, 1985; Kamaya and Higuchi, 1984; Kirk et al., 1986b). The distal His, with the Arg-Leu dipeptide three residues upstream, is conserved.

PROXIMAL HISTIDINE

```
Turnip 153 A V G L S T R D M V A L S G A H T I G Q S R 174
CCP    159 Q R L N M D R E V V A L M G A H A L G K T H 180
HRP    155 V G L N R S S D L V A L S G G H T F G K N Q 176
ML-1   161 A G E F D E K E L V W M L S A H S V A A V N 183
ML-5   161 A G Q F D E L E L V W M L S A H S V A A V N 183
LG-4   162 A G G F D E I E T V W L L S A H S I A A A N 184
LG-5   160 A G E F D E L E L V W M L S A H S V A A A N 182
```

DISTAL HISTIDINE

```
Turnip 31 R M G A S I L R L F F H D C F V N G C D 50
CCP    41 G Y G P V L V R L A W H T S G T W D K H 61
HRP    31 R I A A S I L R L H F H D C F V N G C D 50
ML-1   36 A E A H E S I R L V F H D S I A I S P A 55
ML-5   36 A E A H E S I R L V F H D S I A I S P A 55
LG-4   37 A E A H E A L R M V F H D S I A I S P K 56
LG-5   36 A E A H E S I R L V F H D A I A I S P A 55
```

Figure 3 Comparison of *P. chrysosporium* lignin peroxidase and other peroxidases in the regions of the proximal and distal histidines. Identical amino-acid residues are enclosed in solid boxes. Amino-acid sequences for the turnip, cytochrome c, and horseradish peroxidases were from Welinder and Mazza (1977), Kaput et al. (1982), and Welinder (1976), respectively.

The conservation of His47 and Arg43 would certainly suggest that the mechanism of lignin peroxidase compound I formation is similar to that of other peroxidases. The acid form of the distal His renders the enzyme inactive; thus pH profiles of peroxidase activity (or compound I formation) typically show an inflection at pH 5, corresponding to the pKa of His (see Dunford and Stillman, 1976). However, lignin peroxidase catalysis has a low pH optimum (Tien et al., 1986). Our recent results indicate that the rate of lignin peroxidase compound I formation does not show a pH dependency from pH 2-7.5 (Andrawis et al., 1988). These data suggest that the chemical mechanism of lignin peroxidase compound I formation cannot be identical to the mechanism that was proposed by Poulos and Kraut (1980) for cytochrome c peroxidase, and that is generally accepted for other peroxidases (Manthey and Hager, 1985).

C. Codon Usage

The codon usage of the five lignin peroxidase clones from *P. chrysosporium* is shown in Table 2. A similar table has been published by de Boer et al. (1987), which is derived from the sequences of LG-4 and LG-5. As noted by these workers, the usage is heavily biased toward codons high in C and/ or G residues. With the exception of codons starting with CG and GG (ARg and Gly), the preference of the wobble base within each codons is C > G > T > A. The codons CGG and GGG for Arg and Gly, respectively, are not used at all. de Boer and Kastelein (1986) have noted that biased codon usage occurs extensively with highly expressed genes. The codon usage of the lignin peroxidases is consistent with this observation; the lignin peroxidases are the major secreted proteins produced during secondary metabolism (Tien and Kirk, 1984).

IV. LIGNIN PEROXIDASE GENE STRUCTURE

Evidence has accumulated that the multiple isozymes of lignin peroxidase are encoded by a multigene family. Data from Southern blots and screenings of genomic libraries of *P. chrysosporium* DNA indicate the presence of more than one gene (Tien and Tu, 1987; de Boer et al., 1987; unpublished data from our laboratory). The diversity of the cDNA sequences from the clones isolated so far support this conclusion. Currently a number of laboratories are isolating and characterizing the genes encoding isozymes H2, H6, and H8.

The first reported genomic sequence for the ligninase isozymes was by Smith et al. (1988). We have also isolated the gene that encodes lignin peroxidase isozyme H8 (Holzbaur et al., 1988). A genomic library was constructed from *P. chrysosporium* DNA by partial digestion of the isolated DNA with *Sau*3A to approximately 15-kb fragments followed by ligation to the vector λ-dash (Stratagene). The library was probed with nick-translated cDNA encoding the H8 isozyme. Two distinct genomic clones were isolated and then characterized by Southern blotting. These clones have been sequenced.

A comparison of the cDNA and genomic sequences indicated that both of the clones encode lignin peroxidase isozyme H8. The comparison also showed that the lignin peroxidase gene is split. There are nine coding regions separated by eight introns. The structure of the gene is diagrammed in Figure 4. The exons range in size from 19 to 424 bp, but the introns are similar in length at about 55 bp. This is close to the minimum number of nucleotides in an intron according to current models for splicing. Most

Table 2 Codon Usage Table for Lignin Peroxidase Genes[a]

Amino acid	Codon	Frequency	Amino acid	Codon	Frequency
lys	AAA	2	ala	GCA	23
	AAG	50		GCC	67
asn	AAC	66		GCG	57
	AAT	1		GCT	58
thr	ACA	1	gly	GGA	6
	ACC	59		GGC	96
	ACG	22		GGG	0
	ACT	28		GGT	53
arg	AGA	0	val	GTA	1
	AGG	0		GTC	87
ser	AGC	5		GTC	11
	AGT	2		GTT	16
ile	ATA	0	stop	TAA	4
	ATC	97		TAG	1
	ATT	13	tyr	TAC	1
met	ATG	35		TAT	0
gln	CAA	12	ser	TCA	2
	CAG	99		TCC	49
his	CAC	39		TCG	41
	CAT	2		TCT	17
pro	CCA	7	stop	TGA	0
	CCC	66	cys	TGC	31
	CCG	56		TGT	8
	CCT	27	trp	TGG	15
arg	CGA	1	leu	TTA	0
	CGC	26		TTG	3
	CGG	0	phe	TTC	135
	CGT	23		TTT	4
leu	CTA	0			
	CTC	90			
	CTG	17			
	CTT	17			
glu	GAA	5			
	GAG	80			
asp	GAC	91			
	GAT	30			

[a]Data compiled from a total of 5 lignin peroxidase genes.

5'   3'

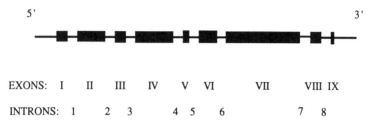

EXONS: I   II   III   IV   V   VI   VII   VIII IX

INTRONS: 1   2   3   4   5   6   7   8

Figure 4   Structure of the *P. chrysosporium* lignin peroxidase isozyme H8 gene. (From Holzbaur et al., 1988, with permission.)

filamentous fungal introns identified to data are short—in the range of 48 to 240 bp (Ballance, 1986). The exon-intron splice junctions share the universal [GT(A/G)ACT. . .(C/T)AG] consensus signal (Mount, 1982). These splice junction sequences are described in Table 3. There is also a putative internal consensus sequence in the intron. The sequence CTGAAC is located 11 to 12 bp from the 3' end of the introns. This internal consensus is similar in sequence and location to that proposed for *Neurospora crassa* genes, which is (A/G)CT(A/G)AC and is located 7 to 19 bp from the 3' end of the intron (Orbach et al., 1986; Germann et al., 1988). These fungal sequences are similar to, although not as strictly conserved as, the *S. cerevisiae* internal consensus sequence of TACTAACA, located 18 to 53 bp from the 3' splice junction (Langford and Gallwitz, 1983). With the exception of intron 5 in the lignin peroxidase genomic clone, splicing

Table 3   Conserved Intronic Sequences

	Intron	5' Site	Internal consensus		3' Site	Reference
Lignin peroxidase H8	1	GTATGC	CTGAAC	- 12 bp -	CAG	
	2	GTGAGT	TTGAAC	- 9 bp -	CAG	
	3	GTAAGT	CTGATC	- 9 bp -	CAG	
	4	GTATAC	CTAACT	- 9 bp -	TAG	
	5	GTAAGT	CTGACT	- 7 bp -	CAG	
	6	GTAAGT	CTGAGAC	- 6 bp -	CAG	
	7	GTGCGT	CTGAAC	- 10 bp -	CAG	
	8	GTGAGT	CTGACA	- 12 bp -	CAG	
P. chrysosporium		GT^A_GAGT	CTGAAC	7-12 bp	^C_TAG	
N. crassa		GTA^A_CGT	^A_GCT^A_GA	7-19 bp	^C_TAG	25, 26
S. cerevisiae		GTATGT	TACTAAC	18-53 bp	^C_TAG	27

occurs at the first AG downstream from the internal consensus sequence, as it does with the *N. crassa* and *S. cerevisiae* introns (Orbach et al., 1986). Both the similarities and the differences in these internal consensus splicing signals may be indicative of species-specific splicing mechanisms. The intra-intronic sequence from *S. cerevisiae* has been shown to be an essential recognition sequence for accurate splicing in vivo (Langford et al., 1984). Woudt et al. (1985) have shown that the *N. crassa* histone H3 gene is correctly spliced in yeast, but at a low efficiency. Confirmation that the *P. chrysosporium* consensus sequence is a required signal for splicing must await the development of a transformation system that will allow the replacement of native genes with specifically mutated or hybrid genes. Several labs are currently focusing on the development of such a transformation system (see below).

The number of introns within filamentous fungal genes is variable. Genomic clones have been sequenced that contain from zero to six introns (Ballance, 1986). The *P. chrysosporium* lignin peroxidase gene is at the top of the range with eight intervening sequences.

The nucleotide sequence of the 5'-flanking DNA for the lignin peroxidase isozyme H8 gene is illustrated in Figure 5. The promoter contains the eukaryotic transcriptional control signals ACAAT at − 104 from the Met initiation codon and a TATAA box 76 bp upstream of the ATG (Breathnach and Chambon, 1981). The TATAA and ACAAT elements have been identified in filamentous fungal genes but the sequences are not always present (Ballance, 1986). These consensus sequences have been shown to be important for the expression of some (Grosveld et al., 1982; Dierks et al., 1983; Charnay et al., 1985) but not all eukaryotic genes (Benoist and Chambon, 1981; Fromm and Berg, 1982). The initiation codon is flanked

C G G C G A $\boxed{\text{A C A A T}}$ A C C A G G A C G T C G A C C A C

G C C T A G G G $\boxed{\text{T A T A A}}$ A A G G G C G $\boxed{\text{A}}$ C A G G A C C

A C C G C A G T C C C T C A G A C A T C C A G T C T C T T

C A G T C C C A C T C A G C A C C A G C A A C A C A G C

G G A C $\boxed{\text{A T G}}$ ...

Figure 5 Promoter sequence from the gene for lignin peroxidase isozyme H8. The consensus promoter elements ACAAT and TATA are indicated, as well as the sites for initiation of transcription and of translation. (From Holzbaur et al., 1988, with permission.)

by the sequence GACATGG, which is homologous to the conserved translation start site in eukaryotes and fungi (Ballance, 1986).

The initiation of transcription downstream from the TATA and ACAAT consensus signals was verified by S1 mapping (Figure 6). A uniformly labeled probe corresponding to the 5' end of the genomic clone was hybridized to mRNA isolated from cultures grown under conditions of nitrogen limitation. After digestion with S1 nuclease, a major protected band of 130 bases was identified on a denaturing polyacrylamide gel. Comparison to a sequencing ladder identified the primary start of transcription as the adenine that is located eight bases 3' to the TATA box. Some size heterogeneity of the protected fragment was observed, which may be an artifact of the S1 nuclease digestion. However, varying the concentration of the S1 nuclease did not produce a single band. Alternatively, there may be multiple start sites for the initiation of transcription of the *P. chrysosporium* gene. S1 mapping of the 5' end of the transcript of the *N. crassa am* gene, which encodes NADP-specific glutamate dehydrogenase, showed at least four initiation sites (Kinnaird and Fincham, 1983). Multiple transcriptional start sites may be typical of the expression of genes in filamentous fungi (Ballance, 1986).

V. REGULATION OF LIGNIN PEROXIDASE EXPRESSION

Numerous studies have been performed on the effects of culture conditions on the levels of lignin peroxidase activity. Cultures of *P. chrysosporium* produce lignin peroxidase during secondary metabolism, upon limitation of the nutrients nitrogen (Keyser et al., 1978), carbon, or sulfur (Jeffries, 1981). Cultures incubated under high oxygen tension showed increased ligninolytic activity (Kirk et al., 1986a: Bar-Lev and Kirk, 1981), and agitated cultures showed a decreased level of activity (Kirk et al., 1986; Ulmer et al., 1983). The observed induction of lignin peroxidase under

---

Figure 6 S1 nuclease mapping analysis of the transcription initiation site. The 2-kb uniformly labeled probe (lane P) was hybridized to 9 μg of *P. chrysosporium* poly A RNA at 51 °C and digested with S1 nuclease at 1000 U/ml. The resulting protected band (lane S) was determined to be 130 bases by comparison with the sequencing ladder (A, G, C, T). The probe was also hybridized to 50 μg of yeast tRNA and digested from S1 nuclease (lane R). (From Holzbaur et al., 1988, with permission.)

nitrogen or carbon limitation is repressed by the addition of L-glutamate but not by α-ketoglutarate (Faison and Kirk, 1985). Lignin is not required for induction of the enzyme, but it has been reported that the addition of lignin to the culture medium increased the observed level of activity two- to fivefold (Faison and Kirk, 1985). Similar increases in the level of activity have also been observed upon the addition of veratryl alcohol, a secondary metabolite of the fungus and an intermediate in the lignin biodegradable pathway. Because lignin itself cannot enter the cell, it has been proposed that veratryl alcohol is the intermediary inducer of increased lignin peroxidase expression in vivo (Leisola et al., 1987; Faison and Kirk, 1985; Faison et al., 1986).

Regulation of lignin peroxidase activity may occur at the level of transcription, translation, or posttranslational modification. Using activity assays and Northern and Western blotting techniques, Tien and Tu (1987) analyzed ligninase isozyme H8 expression over a 6-day time course. It was found that the increase in the level of poly A RNA encoding lignin peroxidase H8 paralleled the increased levels of lignin peroxidase and of veratryl alcohol-oxidation activity. Thus lignin peroxidase expression is regulated at the level of mRNA production.

Lignin peroxidase may also be regulated at the level of posttranslational processing. The enzyme is proteolytically cleaved to remove the signal sequence (Tien and Tu, 1987), glycosylated (Tien and Kirk, 1984), and phosphorylated (M. Tien, unpublished observations). The extent of glycosylation may affect the stability of the secreted proteins. Studies are currently underway in our laboratory to investigate the possible relationship between the phosphorylation state of the enzyme and its activity (I-Ching Kuan and Ming Tien, unpublished observations).

The presence of multiple genes encoding the lignin peroxidase isozymes raises the question of their regulation—are the isozymes coordinately or differentially regulated? The pattern of lignin peroxidase isozymes produced by *P. chrysosporium* is dependent on the composition of the culture medium, the time of enzyme isolation, and the degree of agitation of the cultures during growth. Therefore it is likely that the genes are regulated differentially in response to environmental stress on the fungus.

To address this question directly, we compared the relative levels of expression of lignin peroxidase genes when the fungus was grown under conditions of nitrogen and of carbon limitation. When *P. chrysosporium* was limited for nitrogen, the peak in lignin peroxidase activity occurred at about day 5. The major isozyme produced was H8, although, as can be seen in Figure 7, there were significant concentrations of the isozymes

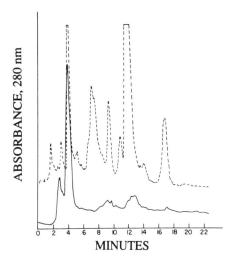

Figure 7 FPLC profile of the lignin peroxidase isozymes from the extracellular fluid of *P. chrysosporium*. The upper, dotted trace is the extracellular fluid from nitrogen-limited cultures. The major peaks eluting at 4 and 12 min correspond to isozymes H2 and H8, respectively. The lower, solid trace is the extracellular fluid from carbon-limited cultures. The major peak eluting at 4 min is isozyme H2. The minor peak eluting at 3 min is isozyme H1. (From Holzbaur et al., 1988, with permission.)

H1, H2, and H6. By day 6, H8 accounts for approximately 1% of the total cellular protein, based on estimates of relative mRNA abundance (Tien and Tu, 1987).

Under conditions of carbon limitation, however, the observed peak in lignin peroxidase activity occurs on day 3 (Faison and Kirk, 1985). Leisola et al. (1987) separated the extracellular peroxidases produced by *P. chrysosporium* by analytical isoelectric focusing. These workers identified seven hemeproteins in the extracellular fluid from carbon-limited cultures and 10 hemeproteins from nitrogen-limited cultures. However, these workers grew their carbon-limited cultures with agitation, whereas their nitrogen-limited cultures were static, thus introducing an additional variable.

In our study the *P. chrysosporium* was grown under the culture conditions of Tien and Kirk (1984), with the exception that the glucose concentration was decreased to 0.1% and the nitrogen was increased to 3 mM using ammonium tartrate. The cultures were grown statically at 39 °C and were harvested at the peak of enzyme activity on day 3. Both the cellular poly A RNA and the proteins in the extracellular fluid were isolated.

Northern blots were run with poly A RNA isolated from both nitrogen- and carbon-limited cultures (Figure 8). The samples were run along with standards of in vitro-transcribed RNA encoding H2 and H8. The H2 probe hybridized to poly A RNA from both nitrogen- and carbon-limited cultures. However, the H8 probe hybridized only to RNA isolated from nitrogen-limited cultures. Therefore, the H2 gene is expressed under conditions of both nitrogen and carbon limitation, but lignin peroxidase isozyme H8 is not expressed in response to carbon limitation for growth.

These results were correlated with the patterns of isozymes produced by the fungus under the two sets of conditions. The FPLC elution profiles are illustrated in Figure 7. Under nitrogen limitation both H2 and H8 are abundant, but under carbon limitation isozyme H8 was not observed.

Therefore, the multiple genes for the lignin peroxidase isozymes are not coordinately regulated in response to environmental stress. Also, it is unlikely that there is a single state of "secondary metabolism" if different genes are transcribed in response to different conditions of nutrient deprivation.

An additional question that must be addressed is the regulation of the complete ligninolytic system. Lignin peroxidase alone is insufficient to completely degrade lignin. The other components of this system, which include the production of extracellular H_2O_2, are also likely to be induced during secondary metabolism. However, are these factors regulated in parallel with lignin peroxidase production, and what is the mechanism of the coupling? Characterization of the components of the ligninolytic system and analyses of the expression of the genes involved should answer these questions.

It remains unclear why there are multiple isozymes for lignin peroxidases. The genes are highly homologous, and the proteins exhibit similar kinetics and substrate specificity (Farrell et al., 1989). Comparison of the promoters of the lignin peroxidase genes may elucidate the mechanisms for the variation in the expression of the isozymes in response to environmental stimuli. More extensive comparisons of the relative conservation of nucleotide homology in the exons and intervening sequences may shed light on the evolution of this multigene family, but more detailed enzymatic characterization will be required in order to functionally differentiate the isoenzymes.

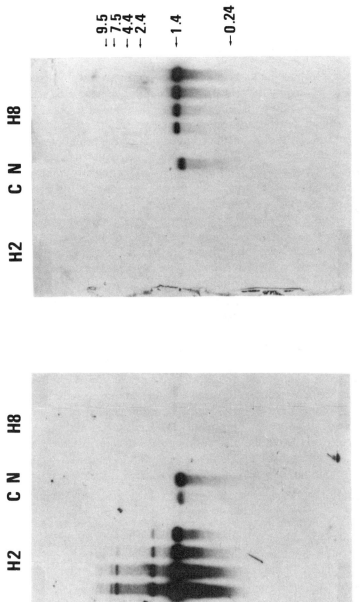

Figure 8 Northern blot of poly A RNA isolated from carbon-limited (C) and from nitrogen-limited (N) cultures. In vitro-transcribed RNA from isozymes H2 (60, 40, 20, and 10 ng) and H8 (10, 20, 40, and 60 ng) were run as internal standards to determine the specificity of the probes. The blot in the left panel was probed with nick-translated H2 cDNA, and the blot in the right panel was probed with nick-translated H8 cDNA. (From Holzbaur et al., 1988, with permission.)

VI. HETEROLOGOUS EXPRESSION OF LIGNIN PEROXIDASE

A. Expression in *E. coli*

The potential environmental and commercial applications of lignin peroxidase have been limited by the relatively low levels of enzyme production possible with current culture techniques. Researchers initially focused on increasing the lignin peroxidase production of *P. chrysosporium* by altering culture parameters such as media composition or the extent of agitation or aeration. Mutants of the fungus were developed with altered growth characteristics. Also, a strategy was developed and is currently employed in our laboratory to screen mutants by growth on lignin-model amino acid adducts (Tien et al., 1987).

With the cloning of the lignin peroxidase cDNAs, work in several laboratories is now focusing on the overexpression of lignin peroxidase in heterologous systems. The first description of the heterologous expression of lignin peroxidase isozyme H8 in *E. coli* was made by Maione et al. (1987). The authors report that their recombinant lignin peroxidase was antigenically cross-reactive with the *P. chrysosporium* protein, had a similar Soret spectrum, and was active in oxidizing veratryl alcohol and lignin-model compounds. In heterologous expression experiments in our laboratory, the recombinant ligninase synthesized by *E. coli* lacked the heme group and was found to be insoluble upon cell rupture. Thus the purification scheme required a refolding step following heme insertion. Farrell and co-workers at Repligen-Sandoz Research Corporation have analyzed the expression of recombinant ligninase in *E. coli* extensively (Farrell et al., 1987). They generated a series of promoter-cDNA constructs that produce antigenically cross-reactive ligninase or ligninase-fusion proteins when transformed into appropriate *E. coli* strains. Following lysis of the recombinant cells, the immunoreactive protein was isolated from the insoluble cell debris by extraction in urea. The ligninase was purified by anion-exchange and gel-filtration chromatography. The recombinant protein was found to lack the heme group. However, the apoprotein could be reconstituted by the addition of excess protoheme IX followed by dialysis in the presence of glutathione. The resulting recombinant holoenzyme was reported to be active in oxidation assays with lignin-model compounds and with lignin.

Successful expression of recombinant ligninase in heterologous cells is, therefore, possible. However, although there are a large number of procedures available for the reconstitution of denatured proteins (Nagai and

Thogersen, 1984; Marston et al., 1984; Nagai et al., 1985), the necessity for protein refolding and heme insertion presents an obstacle for industrial applications. Downstream processing can be very costly and may undermine any potential use of recombinant lignin peroxidases.

B. Expression in Other Heterologous Systems

To circumvent this problem, a number of laboratories are attempting to express the lignin peroxidase in *Aspergillus nidulans*. This is a promising approach, solely because a fungal expression system may be more accepting of a foreign fungal gene. To date, no progress has been reported.

Work is continuing in our laboratory and in the laboratories of other investigators on the expression of lignin peroxidase isozymes in *E. coli*, yeast, and mammalian tissue culture cells. Western blot analyses have indicated that the former two systems produce immunoreactive protein upon induction of the recombinant genes in the appropriate plasmid/promoter constructs. We have expressed lignin peroxidase approaching 10% of total cellular proteins in *E. coli*. Using conventional refolding techniques (similar to that described by Repligen/Sandoz [Farrell et al., 1987]), we have not obtained high-level reconstitution.

VII. TRANSFORMATION SYSTEMS

The first industrial utilization of a lignin-degrading system that is a product of genetic engineering may come from *P. chrysosporium* itself. A large number of potential applications are recognized for the whole fungus: biopulping (Eriksson and Kirk, 1985), degradation of lignin-derived waste effluent (Kirk and Chang, 1981), and degradation of environmental pollutants. Kirk and co-workers are focusing on the use of *P. chrysosporium* for biopulping. Chang and co-workers are using *P. chrysosporium* for degradation of the E1 effluent of the pulping process. Aust and co-workers, who were the first to recognize and demonstrate the use of *P. chrysosporium* for the degradation of environmental pollutants, are presently field testing at chemical dump sites. Genetically engineered *P. chrysosporium* with hyperligninolytic properties would be most desirable for all of the above processes. One approach being used is to place the genes encoding lignin-degrading (or pollutant-degrading) enzymes behind strong promoters. The promoters for the ligninolytic system are turned on only during nutrient limitation. This imposes a limitation on the amount of total ligninolytic activity. It would be more desirable to have activity proportional to cell mass.

The above technologies require the development of a transformation system for *P. chrysosporium*. Transformation systems for fungi are fairly commonplace now (Hynes, 1986). A number of selectable markers involving drug resistance (neomycin and hygromycin) are available. The techniques for the transformation of *P. chrysosporium* are being developed. Gold and co-workers (1983) have developed a system for protoplast formation and regeneration, analogous to other systems developed for fungi. Gold's laboratory has also isolated a large number of amino acid auxotrophs (Gold et al., 1982) and has successfully transformed an adenine auxotroph with the *adeZ* gene of *Schizophyllum commune* (Alic et al., 1989). Reddy's group and our laboratory are also working on the development of stable transformation systems for *P. chrysosporium* based on either complementation of auxotrophic mutants or on transformation to drug resistance. These transformation systems will be valuable tools for the utilization of *P. chrysosporium* throught the development of hyper-producing ligninase strains, as well as for basic studies on the regulation of lignin biodegradation.

VIII. CONCLUDING COMMENTS

Despite the rapid progress in lignin biodegradation the past 5 years, much still remains to be learned about this complicated process. One challenging problem is the determination of the mechanism of degradation of the lignin polymer by isolated ligninases. To date, no studies have demonstrated extensive in-vitro depolymerization. The value of the application of molecular biology to this problem is quite clear. However, it is just a tool; to truly understand how lignin is degraded, we must also rely on the expertise of a nearly extinct discipline—lignin chemistry. Future breakthroughs, like past ones, will be dependent on the interaction of this discipline with others, and indeed, molecular biology will play a key role. Thus, as in the past and for the forseeable future, characterizing lignin biodegradation will be a multidisciplinary endeavor.

REFERENCES

Adler, E. (1977). Lignin chemistry. Past, present and future. *Wood Sci. Technol.* *11*:169-218.
Alic, M., Kornegay, J.R., Pribnow, D., and Gold, M.H. (1989). Transformation by complementation of an adenine auxotroph of the lignin-degrading basidiomycete Phanerochaete chrysosporium. *Appl. Environ. Microbiol.* *55*:406-411.

Andrawis, A., Johnson, K.A., and Tien, M. (1988). Studies on compound I formation of the lignin peroxidase from *Phanerochaete chrysosporium*. *J. Biol. Chem. 263*:1195-1198.

Ballance, D.J. (1986). Sequences important for gene expression in filamentous fungi. *Yeast 2*:229-236.

Bar-Lev, S.S. and Kirk, T.K. (1981). Effects of molecular oxygen on lignin degradation by *Phanerochaete chrysosporium*. *Biochem. Biophys. Res. Comm. 99*:373-378.

Bell, G.I., Santerre, R.F., and Mullenbach, G.T. (1983). Hamster preproglucagon contains the sequence of glucagon and related peptides. *Nature 302*:716-718.

Benoist, C. and Chambon, P. (1981). In vivo sequence requirements of the SV40 early promoter region. *Nature 290*:304-310.

Breathnach, R. and Chambon, P. (1981). Organization and expression of eukaryotic split genes coding for proteins. *Am. Rev. Biochem. 50*:349-383.

Charnay, P., Mellon, P., and Maniatis, T. (1985). Linker scanning mutagenesis of the 5'-flanking region of the mouse β-major-globin gene: sequence requirements for transcription in erythroid and nonerythroid cells. *Mol. Cell Biol. 5*:1498-1511.

de Boer, H.A. and Kastelein, R. (1986). Biased codon usage: an exploration of its role in optimization of translation. In *Maximizing Gene Expression*, Reznikoff, W. and Gold, L., eds., Butterworth, Boston, p. 225-286.

de Boer, H.A., Zhang, Y.Z., Collins, C., and Reddy, C.A. (1987). Analysis of nucleotide sequences of two ligninase cDNAs from a white-rot filamentous fungus, *Phanerochaete chrysosporium*. *Gene 60*:93-102.

Dierks, P., van Ooyen, A., Cochran, M.D., Dobkin, C., Reiser, J., and Weissman, C. (1983). Three regions upstream from the cap site are required for efficient and accurate transcription of the rabbit β-globin gene in mouse 3T6 cells. *Cell 32*:695-706.

Dunford, H.B. and Stillman, J.S. (1976). On the function and mechanism of action of peroxidases. *Coord. Chem. Rev. 19*:187-251.

Edge, A.S., Faltynek, C.R., Hof, L., Reichert, L.E. Jr., and Weber, P. (1981). Deglycosylation of glycoproteins by trifluoromethane sulfonic acid. *Anal. Biochem. 118*:131-137.

Eriksson, K.-E. and Kirk, T.K. (1985). Biopulping, biobleaching and treatment of kraft bleaching effluents with white-rot fungi. In *Comprehensive Biotechnology,* Vol. 3. Robinson, C.W., ed., Pergamon Press, Toronto.

Faison, B.D. and Kirk, T.K. (1985). Factors involved in the regulation of a ligninase activity in *Phanerochaete chrysosporium*. *Appl. Environ. Microbiol. 49*:299-304.

Faison, B.D., Kirk, T.K., and Farrell, R.L. (1986). Role of veratryl alcohol in regulating ligninase activity in *Phanerochaete chrysosporium*. *Appl. Environ. Microb. 52*:251-254.

Farrell, R.L., Gelep, P., Anilionis, A., Javaherian, K., Maione, T.E., Rusche, J.R., Sadownick, B.A., and Jackson, J.A. (1987). European Patent Application #87810516.2.

Farrell, R.L., Murtagh, K.E., Tien, M., Mozuch, M.D., and Kirk, T.K. (1989). *Enzyme Microb. Technol. 11*:322-328.

Fromm, M. and Berg, P. (1982). Deletion mapping of DNA regions required for SV40 early region promoter function in vivo. *J. Mol. Appl. Genet. 1*:457-481.

Germann, U.A., Miller, G., Hunziker, P.E., and Lerch, K. (1988). Characterization of two allelic forms of *Neurospora crassa* lactase amino- and carboxyl-terminal processing a precursor. *J. Biol. Chem. 263*:885-896.

Glenn, J.K. and Gold, M.H. (1985). Purification and characterization of an extracellular Mn(II)-dependent peroxidase from the lignin-degrading basidiomycete *Phanerochaete chrysosporium. Arch. Biochem. Biophys. 242*:329-341.

Glenn, J.K., Morgan, M.A., Mayfield, M.B., Kumahara, M., and Gold, M.H. (1983). An H_2O_2-requiring enzyme preparation involved in lignin biodegradation by the white rot basidiomycete *Phanerochaete chrysosporium. Biochem. Biophys. Res. Comm. 114*:1077-1083.

Gold, M.H. (1988). The Third Chemical Congress of North America. *Biotechnology 24*.

Gold, M.H., Cheng, T.M., and Mayfield, M.B. (1982). Isolation and complementation studies of auxotrophic mutants of the lignin-degrading basiciomycete *Phanerochaete chrysosporium. Appl. Environ. Microb. 44*:996-1000.

Gold, M.H., Cheng, T.M., and Alic, M. (1983). Formation, fusion, and regeneration of protoplasts from wild-type and auxotrophic strains of the white-rot basidiomycete *Phanerochaete chrysosporium. Appl. Environ. Microb. 46*:260-263.

Goring, D.A.I. (1977) A speculative picture of the delignification process. In *Cellulose Chemistry and Technology, ACS Symposium Series No. 48,* Arthur, J.C. Jr., ed., Am. Chem. Soc., Washington, D.C., p. 273-277.

Grosveld, G.C., Rosenthal, A., and Flavell, R.A. (1982). Sequence requirements for the transcription of the rabbit β-globin gene in vivo: the − 80 region. *Nucleic. Acids Res. 10*:4951-4971.

Higuchi, T. (1985). Degradative pathways of lignin model compounds. In *Biosynthesis and Biodegradation of Wood Components.* Academic Press, Orlando, FL, p. 557-558.

Holzbaur, E.L.F., Andrawis, A., and Tien, M. (1988). Structure and regulation of a lignin peroxidase gene from *Phanerochaete chrysosporium. Biochem. Biophys. Res. Comm. 155*:626-633.

Hynes, M.J. (1986). Transformation of filamentous fungi. *Experimental Mycology 10*:1-8.

Jeffries, T.W., Choi, S., and Kirk, T.K. (1981). Nutritional regulation of lignin degradation by *Phanerochaete chrysosporium. Appl. Environ. Microbiol. 42*:290-296.

Julius, D., Schekman, R., and Thorner, J. (1984). Glycosylation and processing of prepro-α-factor through the yeast secretory pathway. *Cell 36*:309-318.

Kamaya, T. and Higuchi, T. (1984). Metabolism of non-phenolic diarylpropane lignin substructure model compound by *Coriolus versicolor. FEMS Microbiol. Lett. 2*:89-92.

Kaput, J., Goltz, S., and Blobel, G.J. (1982). Nucleotide sequence of the yeast nuclear gene for cytochrome c peroxidase precursor. *J. Biol. Chem. 257*:15054-15058.

Kawai, S., Umezawa, T., and Higuchi, T. (1985). Arylglycerol-γ-formyl ester as an aromatic ring cleavage product of nonphenolic β-O-4 lignin substructure model compounds degraded by *Coriolus versicolor. Appl. Environ. Microb. 50*:1505-1508.

Keyser, P., Kirk, T.K., and Zeikus, J.G. (1978). Ligninolytic enzyme system of *Phanerochaete chrysosporium*: synthesized in the absence of lignin in response to nitrogen starvation. *J. Bacteriol. 135*:790-797.

Kinnaird, J.H. and Fincham, J.R.S. (1983). The complete nucleotide sequence of the *Neurospora crassa am* (NADP-specific glutamate dehydrogenase) gene. *Gene 26*:253-260.

Kirk, T.K. (1983). Degradation and conversion of lignocellulose. In *The Filamentous Fungi*, Vol. 4, *Fungal Technology,* Smith, J.E., Berry, D.R., and Kristiansen, B., eds., Edward Arnold, London, p. 266-295.

Kirk, T.K. and Chang, H.-M. (1981). Potential applications of biologininolytic systems. *Enzyme Microbiol. Technol. 3*:189-196.

Kirk, T.K., Croan, S., Tien, M., Murtagh, K.E., and Farrell, R.L. (1986a). Production of multiple ligninases by *Phanerochaete chrysosporium*: Effect of selected growth conditions and use of a mutant strain. *Enzyme Microbiol. Technol. 8*:27-32.

Kirk, T.K., Tien, M., Kersten, P.J., Mozuch, M.D., and Kalyanaraman, B. (1986b). Ligninase of *Phanerochaete chrysosporium*. Mechanism of its degradation of the non-phenolic aryl-glycerol β-aryl ether substructure of lignin. *Biochem. J. 236*:279-287.

Kuila, D., Tien, M., Fee, J.A., and Ondrias, M.R. (1986). Resonance raman spectra of extracellular ligninase: evidence for a heme active site similar to those of peroxidases. *Biochemistry 24*:3394-3397.

Langford, C.J. and Gallwitz, D. (1983). Evidence for an intron-contained sequence required for the splicing of yeast RNA polymerase II transcripts. *Cell 33*:519-527.

Langford, C.J., Klinz, F.-J., Donath, C., and Gallwitz, D. (1984). Point mutations identify the conserved, intron-contained TACTAAC box as an essential splicing signal sequence in yeast. *Cell 36*:645-653.

Leisola, M.S.A., Kozulic, B., Meussdoerffer, F., and Fiechter, A. (1987). Homology among multiple extracellular peroxidases from *Phanerochaete chrysosporium*. *J. Biol. Chem. 262*:419-424.

Maione, T.E., Javaherian, K., Belew, M.A., Gomez, L.E., and Farrell, R.L. (1987). In *Lignin Enzymatic and Microbial Degradation*, Odier, E., ed., INRA, Paris, p. 117-183.

Manthey, J.A. and Hager, L.P. (1985). Characterization of the oxidized states of bromoperoxidase. *J. Biol. Chem. 260*:9654-9659.

Marston, F.A.O., Lowe, P.A., Doel, M.T., Schoemaker, J.M., White, S., and Angal, S. (1984). Purification of calf prochymosin (prorennin) synthesized in *Escherichia coli. Biotechnology 2*:800-804.

Mount, S.M. (1982). A catalogue of splice junction sequences. *Nucleic. Acids Res. 10*:459-472.

Nagai, K., Perutz, M.F., and Poyart, C. (1985). Oxygen binding properties of human mutant hemoglobins synthesized in *Escherichia coli. Proc. Natl. Acad. Sci. USA 82*:7252-7255.

Nagai, K. and Thogersen, H.C. (1984). Generation of β-globin by sequence-specific proteolysis of a hybrid protein produced in *Escherichia coli. Nature 309*: 810-812.

Neuberger, A., Gottshalk, A., Marshal, R.D., and Spiro, R.D. (1972). In *The Glycoproteins: Their Composition, Structure and Function Part A,* Gottschalk, A., ed., Elsevier, Amsterdam, p. 450-490.

Orbach, M.J., Porro, E.B., and Yanofsky, C. (1986). Cloning and characterization of the gene for β-tubulin from a benomyl-resistant mutant of *Neurospora crassa* and its use as a dominant selectable marker. *Mol. Cell. Biol. 6*: 2452-2461.

Poulos, T.L. and Kraut, J. (1980). The stereochemistry of peroxidase catalysis. *J. Biol. Chem. 255*:8199-8205.

Proudfoot, N.J. and Brownbee, G.G. (1976). 3' Non-coding region sequences in eukaryotic messenger RNA. *Nature 263*:211-214.

Smith, T.L., Schalch, H., Gaskell, J., Covert, S., and Cullen, D. (1988). Nucleotide sequence of a ligninase gene from *Phanerochaete chrysosporium. Nucleic Acids Res. 16*:1219.

Tien, M. and Kirk, T.K. (1983). Lignin-degrading enzyme from hymenomycete *Phanerochaete chrysosporium. Science 221*:661-663.

Tien, M. and Kirk, T.K. (1984). Lignin-degrading enzyme from *Phanerochaete chrysosporium.* Purification, characterization, and catalytic properties of a unique H_2O_2-requiring oxygenase. *Proc. Natl. Acad. Sci. USA 81*:2280-2284.

Tien, M., Kirk, T.K., Bull, C., and Fee, J.A. (1986). Steady-state and transient-state kinetic studies on the oxidation of 3,4-dimethoxybenzyl alcohol catalyzed by the ligninase of *Phanerochaete chrysosporium. J. Biol. Chem. 261*: 1687-1693.

Tien, M. and Tu, C.-P.D. (1987). Cloning and sequencing of a cDNA for a ligninase from *Phanerochaete chrysosporium. Nature 326*:520-523.

Tien, M., Kersten, P.J., and Kirk, T.K. (1987). Selection and improvement of lignin-degrading microorganism: potential strategy based on lignin model-amino acid adducts. *Appl. Environ. Microbiol. 53*:242-245.

Ulmer, D.C., Leisola, M., Puhakka, J., and Fiechter, A. (1983). *Phanerochaete chrysosporium:* growth pattern and lignin degradation. *Eur. J. Appl. Microbiol. Biotechnol. 18*:153-157.

Umezawa, T. and Higuchi, T. (1985). Aromatic ring cleavage in degradation of β-O-4 lignin substructure by *Phanerochaete chrysosporium. FEBS Lett. 182*: 257-259.

Welinder, K.G. (1976). Covalent structure of the glycoprotein horseradish peroxidase (EC1.11.17). *FEBS Lett. 72*:19-23.

Welinder, K.G. and Mazza, G. (1977). Amino-acid sequences of heme-linked, histidine-containing peptides of five peroxidases from horseradish and turnip. *Eur. J. Biochem. 73*:353-358.

Woudt, L.P., van den Heurel, J., van Raamsdonk-Duin, M.M.C., Mager, W.H., and Planta, R.J. (1985). Correct removal by splicing of a *Neurospora* intron in yeast. *Nucleic Acids Res. 13*:7729-7739.

Zhang, Y.Z., Zylstra, G.J., Olsen, R.H., and Reddy, C.A. (1986). Identification of cDNA clones for ligninase from *Phanerochaete chrysosporium* using synthetic oligonucleotide probes. *Biochem. Biophys. Res. Comm. 137*:649-656.

9

Genetics and Biochemistry of Toxin Synthesis in *Cochliobolus* (*Helminthosporium*)

Jonathan D. Walton
Michigan State University, East Lansing, Michigan

I. INTRODUCTION AND TAXONOMY

Helminthosporium is the most commonly used name for a large genus of filamentous fungi that produce similar conidia (asexual spores). In the most recent taxonomic revision, Alcorn (1983) has split the genus into *Bipolaris, Drechslera,* and *Exserohilum.* Each of these genera has a unique perfect state: *Cochliobolus, Pyrenophora,* and *Setosphaeria,* respectively. *Cochliobolus* is the name for the sexual stage (teleomorph) of the subgroup *Bipolaris* and is the subject of this review. However, because of its long history, familiarity, and continued use by many workers in the field of plant pathology, the name *Helminthosporium* is also used in this chapter.

Like many filamentous fungi, species of *Cochliobolus* produce manifold secondary metabolites as well as many extracellular enzymes such as proteases and hydrolases. Interesting secondary metabolites include the cytochalasins, ophiobolins, helminthosporal, and helminthosporol (Hesseltine et al., 1971).

Many species in *Helminthosporium, sensu latu,* including those now more properly referred to as *Cochliobolus,* can grow as saprophytes on dead organic plant material. Some species are economically important because they are also capable of parasitizing plants, and it is as plant pathogens of graminaceous crops that this group of fungi has received the most attention. Two severe plant epidemics, the Victoria blight of oats in the 1940s and the Southern Corn Leaf Blight epidemic in 1970, were due to *Cochliobolus.* The former was caused by *C. victoriae* (= *Bipolaris victoriae* or *Helminthosporium victoriae*) and the latter by *C. heterostrophus* race T (= *B. maydis* or *H. maydis* race T). Other species that cause chronic disease losses include *C. miyabeanus* (= *B. oryzae* or *H. oryzae*), *C. sativus* (= *B. sorokiniana* or *H. sativum*), *C. carbonum* (= *B. zeicola* or *H. carbonum*), *Pyrenophora tritici-repentis* (= *Drechslera tritici-repentis* or *H. tritici-repentis*), and *Setosphaeria turcicum* (= *Exserohilum turcicum* or *H. turcicum*).

II. HOST-SELECTIVE PHYTOTOXINS FROM *HELMINTHOSPORIUM*

Four and perhaps six species of *Helminthosporium* produce what are known as *host-specific* or *host-selective* phytotoxins. Five have *Cochliobolus* and one has *Pyrenophora* as the perfect stage. Criteria that must be satisfied for a compound to be considered as such include: (a) an absolute correlation between the ability to produce the compound and the pathogenicity or virulence of the producing fungus, (b) an absolute correlation between the sensitivity to the compound and susceptibility to the producing fungus, and (c) production of the compound by the fungus during the appropriate stages of pathogenesis (Scheffer and Briggs, 1981; Scheffer and Livingston, 1984; Scheffer, 1983). Host-selective toxins have been called *disease determinants,* referring to their critical role in determining the outcome (i.e., either susceptibility or resistance) of the interaction between a fungus and a host plant.

A. Victorin or HV-Toxin

In the 1940s a previously unrecognized fungus, called *H. victoriae,* caused severe losses of oats in the United States. The susceptible varieties were those that had a newly introduced gene for resistance to the known races of the crown rust pathogen, *Puccinia coronata.* The genetic linkage between these two traits is very tight, and they may, in fact, be the same gene. Recent evidence from tissue culture experiments is consistent with a single genetic locus controlling both phenotypes (Rines and Luke, 1985).

Shortly after the discovery of this new pathogen, it was observed that *H. victoriae* culture filtrates had selective activity against the same oat varieties that were susceptible to the fungus. Considerable data has accumulated on the chemistry and mode of action of victorin. The structures of five active victorin species were elucidated in 1985. They are all partially cyclic, chlorinated pentapeptides. The structures of four members of the victorin family are shown in Figure 1; the fifth, called *victoricine*, contains unmodified leucine in place of chloroleucine and hydroxytyrosine (2,5 DOPA) in place of victalanine (Wolpert et al., 1985, 1986; Gloer et al., 1985; Kono et al., 1986).

Victorin is probably the most phytotoxic compound known. It inhibits root growth half-maximally at 100 pM, yet resistant oats as well as all other plants tolerate concentrations 10^5-10^6 times higher (Walton and Earle, 1984). Victorin inhibits many cellular processes (Daly, 1981). It stimulates sensitive oat protoplasts to secrete a glucan polysaccharide (Walton and Earle, 1985). Wolpert and Macko (1989) have recently reported that victorin derivatized with radioiodinated Bolton-Hunter reagent binds selectively and covalently to a 100-kDa microsomal protein in susceptible but not resistant oats. The binding is selective only in vivo (i.e., in leaf slices). Binding occurs but is nonselective in vitro.

B. T-Toxin

In a situation similar to that of *H. victoriae* and oats, the appearance of *H. maydis* (*C. heterostrophus*) race T as a major pathogen of maize was

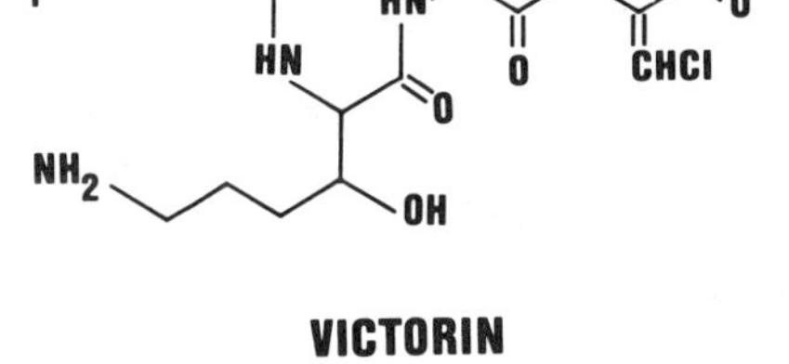

	R_1	R_2
Victorin B	CH_2Cl	OH
C	$CHCl_2$	OH
D	$CHCl_2$	H
E	CCl_3	OH

VICTORIN

Figure 1 Structure of victorin from *Cochliobolus victoriae*. For structure of victoricine see test. (From Wolpert et al., 1985, 1986.)

due to the widespread introduction into maize of a genetic trait called T-cms, for Texas cytoplasmic male sterility. Unfortunately, plants containing T-cms are also very sensitive to a compound, called T-toxin, produced by *H. maydis* race T, and under the favorable environmental conditions of 1970 some 15% of the U.S. maize crop was destroyed.

T-toxin comprises a family of polyketols, all linear hydrocarbons substituted with a variety of hydroxyl and carbonyl functionalities. The major toxin species contain 37-43 carbon atoms (Kono and Daly, 1979). Isolates of *H. maydis* that do not show selectivity on T-cms maize (designated race 0) do not produce these compounds (Tegtmeier et al., 1982). Synthetic analogs of T-toxin are highly host selective on T-cms maize (Suzuki et al., 1982a, 1982b; Kraus et al., 1984). Sugawara et al. (1987) have recently reported that one of the ophiobolins (a group of well-characterized phytotoxic sesterterpenoids produced by many species of *Helminthosporium*) has host-selective activity against T-cms maize. Evidence for selective activity of this ophiobolin, called 6-epiophiobolin, was found in only one of several bioassays tested. It was active against T-cms maize at 800 nM, and its selectivity (the ratio of activity against T-cms maize vs. maize with normal (N) cytoplasm) was 80-fold. These results are in contrast to the T-toxin characterized by Daly and co-workers, which is active against T-cms maize in many bioassays, is active against T-cms maize at 1-10 nM, and shows activity against N cytoplasm maize only at concentrations 10,000 times higher (Suzuki et al., 1982b). Based on the available data, it seems unlikely that the ophiobolins studied by Sugawara et al. (1987) are specifically involved in the *H. maydis*/T-cms maize interaction.

The molecular biology of the T-cms trait and its relation to susceptibility to *H. maydis* race T and T-toxin has been extensively studied. Only the major results are summarized here. Miller and Koeppe (1971) showed that mitochondria from T-cms maize are selectively sensitive to T-toxin. The mitochondrion has been confirmed to be the site of action of T-toxin by ultrastructural, biochemical, and physiological studies (Aldrich et al., 1977; Matthews et al., 1979; Walton et al., 1979). Forde et al. (1978) and Forde and Leaver (1980) found that mitochondria from T-cms but not N maize synthesize in vitro a novel polypeptide of molecular weight 13,000. Dewey et al. (1986) identified a gene unique to T-cms mitochondria that contains an open reading frame, called URF-13, encoding a 13,000 Da polypeptide. This URF was the result of a rearrangement of the mitochondrial genome unique to T-cms. Mitochondrial DNA from plants regenerated from T-cms maize tissue culture that have reverted to fertility (and to T-toxin insensitivity) lose URF-13 (Gengenbach et al., 1981; Rottmann et al., 1987; Wise et al., 1987), supporting the conclusion that this open reading frame is involved in the sensitivity to T-toxin and hence in susceptibility to *H. maydis* race T.

Most recently, Dewey et al. (1988) have expressed this maize mitochondrial gene, URF-13, in *E. coli*. The bacteria, which are normally insensitive to T-toxin, became sensitive. These data clearly demonstrate that the sensitivity of T-cms maize to *H. maydis* race T is due to the product of URF-13. It is still not known how this 13,000 Da polypeptide confers sensitiviity to T-toxin (Braun et al., 1989).

C. HS-Toxin (Helminthosporoside)

HS-toxin is a family of compounds produced by *Helminthosporium sacchari*, a pathogen of certain cultivars of sugarcane. All of the HS-toxins contain a sesquiterpenoid in one of three isomeric forms. To these sesquiterpenoids are attached, via beta linkages, zero to four galactofuranosyl residues (Livingston and Scheffer, 1981, 1984a; Macko et al., 1981, 1983; Macko, 1983). Those with four galactose residues, two being attached to each side of the central sesquiterpenoid, have toxic activity. The lower homologs (i.e., HS-toxins containing fewer than four galactose residues) can in some cases protect sensitive sugarcane cultivars from the active HS-toxins (Livingston and Scheffer, 1983, 1984b; Duvick et al., 1984). Culture filtrates and mycelia of *H. sacchari* contain an enzyme, beta-galactofuranosidase, that can convert HS-toxin to the lower homologs (Livingston and Scheffer, 1983; Nakajima and Scheffer, 1987). However, the levels of this enzyme in *H. sacchari* might not be sufficient to significantly affect the accumulation of HS-toxin in culture. Other species of *Helminthosporium* produce much more of this enzyme (Livingston and Scheffer, 1983).

Coculturing *H. sacchari* with sugarcane suspension cultures stimulates HS-toxin production. Cultures from resistant sugarcane stimulate as effectively as those from susceptible sugarcane (Larkin and Scowcroft, 1981). The chemical basis of this effect is unknown.

D. HC-Toxin

Race 1 of *Helminthosporium carbonum* is defined by its selective pathogenicity on certain maize cultivars that are homozygous recessive at the nuclear *Hm* locus (Ullstrup and Brunson, 1947). Resistance to races 2 and 3 is polygenic. The selective pathogenicity of race 1 is due to production of a compound called *HC-toxin*, which in vitro has the same specificity as the fungus (Scheffer and Ullstrup, 1965). HC-toxin is a cyclic tetrapeptide of structure cyclo(D-Pro-L-Ala-D-Ala-L-AOE), where AOE stands for 2-amino-8-oxo-9,10-epoxidecanoic acid (Figure 2) (Liesch et al., 1982; Walton et al., 1982; Pope et al., 1983; Kawai and Rich, 1983). The epoxide and the adjacent ketone are required for activity (Ciuffetti et al., 1983; Walton and Earle, 1983; Kim et al., 1987). Analogs of HC-toxin containing glycine or hydroxyproline instead of alanine or proline are also present in minor amounts in culture filtrates of *H. carbonum* race 1 (Kim et al., 1985; Rasmussen and Scheffer, 1988).

Very little is known about the mode of action of HC-toxin. Under some conditions isolated leaf mesophyll protoplasts survive better in the presence than in the absence of HC-toxin (Earle and Gracen, 1982; E.D. Earle, personal communication). HC-toxin stimulates the uptake by roots of several ions and solutes, especially nitrate (Yoder and Scheffer, 1973a, 1973b; see also Lin and Kauer, 1985).

E. Other *Helminthosporium* Host-Selective Toxins

C. miyabeanus, a pathogen of rice, has been reported to produce a host-selective toxin (Vidhyasekaran et al., 1986). On a three-point scale of very severe, moderate, and no effect, there was an absolute correlation between the host range of the fungus and sensitivity to the toxin among 11 species of plants. The compound responsible has not yet been identified, but is

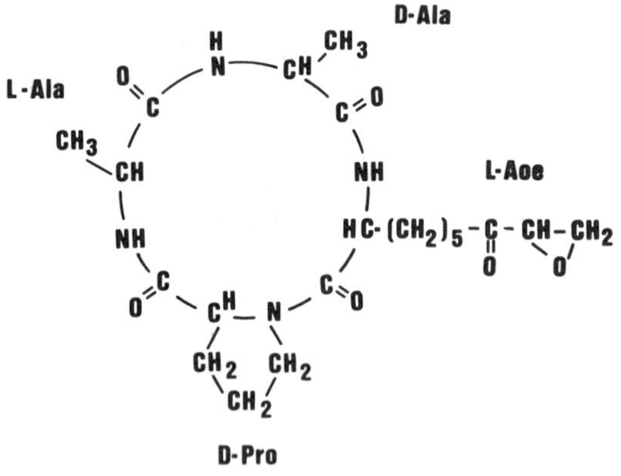

Figure 2 Structure of HC-toxin from *Cochliobolus carbonum* race 1. (From Liesch et al., 1982; Walton et al., 1982; Pope et al., 1983.)

apparently not an ophiobolin. *P. tritici-repentis* produces an uncharacterized toxic compound whose activity against wheat cultivars correlates with susceptibility to the fungus (Tomas and Bockus, 1987).

III. GENETICS OF TOXIN BIOSYNTHESIS

Although *Helminthosporium* and *Bipolaris* are names frequently used for the group of fungi under consideration, *Cochliobolus* is used when referring to the perfect stage of these fungi (Yoder et al., 1986).

A. Classical

Cochliobolus is an ascomycete in the Pyrenomycetes. Like other Ascomycetes, the normal mycelial stage is haploid. *Cochliobolus* produces ascospores in asci, with eight ascospores per ascus. The asci are grouped in larger sacs known as perithecia. The basic genetics of *Cochliobolus* are similar to *Neurospora*, but because the ascospores are twisted around each other, ordered tetrads cannot be recovered from the former. *C. heterostrophus* (= *H. maydis*) is thought to contain eight chromosomes (Guzman et al., 1982). *Cochliobolus* is bipolar heterothallic, meaning that there are two mating types (formerly called *A* and *a*, and now called *MAT1-1* and *MAT1-2*). Successful mating occurs only between two isolates of different mating type.

Mating type is controlled by a nuclear gene with two known alleles. One of the mating loci has recently been cloned (Turgeon et al., 1988). A cosmid containing a cloned mating type locus was identified by introducing a cosmid library containing total genomic DNA from an isolate of mating type MAT1-1 into an isolate of mating type *MAT1-2*. The transformant of interest was identified by its ability to produce fertile perithecia when crossed with the untransformed recipient isolate. The transformant retained the ability to cross with isolates of genotype *MAT1-1* and could also mate with itself. In other words, the transformant was homothallic.

Sexual crosses are made with *Cochliobolus* by placing agar plugs of two isolates of different mating types on opposite sides of an autoclaved maize leaf in a petri plate. The maize leaf works best if it is naturally senescent tissue (Luttrell, 1958; Leach et al., 1982a). When the two mycelia meet on the leaf, ascogenous tissue and then perithecia are formed. If one of the two isolates has a gene (*alb*) for lack of hyphal pigmentation (*Cochliobolus* hyphae are normally grey or black), then the maternal parent in a cross can be determined, since the perithecium is derived from the maternal parent (e.g., Garber and Yoder, 1984).

A basic groundwork has been laid for classical genetic studies with *Cochliobolus*, including methods for growing the fungus on defined medium, generating nutritional and morphological mutants, inducing heterokaryosis and the parasexual cycle, making crosses, isolating ascospores, and storing isolates (Luttrell, 1958; Tinline, 1962; Leach et al., 1982a). A number of single-gene mutants have been described.

Naturally occurring genetic variation of some traits has been studied. Nelson and Sherwood (1968) studied genetic control of endopolygalacturonase production in *C. carbonum*. Kline and Nelson (1969) studied monogenic and polygenic control of pathogenicity of the same fungus on six different grass hosts.

Because of their importance in the process of pathogenesis, the genes in *Cochliobolus* that control production of the host-selective toxins have received the most attention. Even before HC-toxin was discovered, it was demonstrated that the selective pathogenicity of *C. carbonum* was controlled by a single genetic locus (Nelson and Ullstrup, 1961; Nelson and Kline, 1963). Later, it was shown that this "single-gene" control of selective pathogenicity was correlated with the production of HC-toxin (Scheffer et al., 1967). In other words, the production of HC-toxin was controlled by a single genetic locus in *C. carbonum*. In crosses between *C. carbonum* and *C. victoriae* (which makes the oat-selective toxin victorin), pathogenicity on oats, corn, both plants, or neither plant segregated 1:1:1:1, and correlated absolutely with the production of victorin, HC-toxin, both toxins, or neither toxin (Scheffer et al., 1967). The reasonable conclusion is that two single but independent genetic loci control the production of victorin and HC-toxin.

The genetics of T-toxin production by *C. heterostrophus* race T have been intensively studied. Again, the trait segregates as a single genetic locus in crosses between toxin-producing strains (race T) and non-toxin-producing strains (race 0) (Lim and Hooker, 1971; Yoder and Gracen, 1975; Yoder, 1980; Tegtmeier et al., 1982). Analysis of hundreds of crosses has failed to identify more than a single genetic locus controlling T-toxin production. This locus has been named *Tox1* (Leach et al., 1982b).

In the time since much of this genetic work was done, the structures of many of the host-selective toxins from *Helminthosporium* species have been elucidated. Structurally they do not have much in common. They are not simple compounds, nor are they closely related to intermediates in primary metabolism. Most complex secondary products require the participation of many different enzymes encoded by many different genes. How then are we to explain the genetic data indicating that production of the various host-selective toxins is controlled by single (but different) loci?

There are several plausible explanations. Perhaps there are additional genes yet to be discovered. However, if this were the case, then the isolates of *C. victoriae* and *C. carbonum* studied by Scheffer et al. (1967) must have fortuitously shared all but one of the necessary genes to synthesize the other toxin, which seems unlikely. Another possibility is that the *Tox* loci represent mutations in other pathways, and that the toxins are metabolic intermediates that only accumulate to phytotoxic levels in the *Tox* "mutants" (Yoder, 1980). The complex structures of the toxins argues against them being intermediates in the pathways of other metabolites. Moreover, it has been demonstrated that T-toxin is produced by heterokaryons formed between Tox^+ and Tox^- strains of *C. heterostrophus* (Leach et al., 1982b). Thus, production of T-toxin is dominant to nonproduction. If toxin "production" were due to a block in a metabolic pathway, heterokaryons should not accumulate T-toxin.

A third possibility is that each host-selective phytotoxin is biosynthesized by a single multifunctional enzyme/polypeptide. Hence a single gene might encode a single polypeptide with the capability of assembling an entire toxin from basic primary metabolites (Panopoulos et al., 1984; Walton, 1987). Enniatin B, a cyclic hexidepsipeptide from *Fusarium oxysporium*, is synthesized from valine and hydroxyisovaleric acid by a single polypeptide with a molecular weight of 250,000 (Zocher et al., 1982). This enzyme could be a model for how the cyclic peptide host-selective toxins victorin and HC-toxin are made. The polyketide 6-methylsalicylic acid is synthesized by an enzymatic complex composed of identical subunits (Dimroth et al., 1976; Friedrich, 1977). T-toxin is a polyketide. All the cyclic peptide synthetases and 6-methylsalicylic acid synthetase have functional and structural similarities to fatty acid synthetase, which in filamentous fungi is composed of just two different polypeptides (Elovson, 1975; Schweizer et al., 1986).

A fourth possibility is that the *Tox* loci are complex, that is, they are composed of contiguous genes, each of which encodes one of the enzymes in the toxin biosynthetic pathway. This hypothesis is tenable because classical genetic analyses cannot easily resolve tightly linked genes. It is, furthermore, supported by the biochemical data on the biosynthesis of HC-toxin from *Cochliobolus carbonum* (see Section IV).

B. Molecular Genetics

The molecular genetic methods for *Cochliobolus* are well developed, due in large part to the work of Yoder and co-workers at Cornell on *C. heterostrophus*. Techniques for the isolation and fractionation of DNA have been developed (Garber and Yoder, 1983; Yoder, 1988) and mitochondrial plasmids have been described (Garber et al., 1984). Wild-type *C. heterostrophus* has been transformed using the *amdS* gene from *Aspergillus nidulans* (Turgeon et al., 1985). The *amdS* gene encodes the enzyme acetamidase, whose presence allows *A. nidulans* and *C. heterostrophus* to grow on acetamide as the sole source of nitrogen. Transformation occurs by random integration into the chromosomal DNA. Transformation frequencies are low, perhaps because the transforming plasmid has no homology to *Cochliobolus* DNA.

A transformation system based on drug resistance has been developed recently (Turgeon et al., 1987). The selectable marker, called *hygB*, encodes an enzyme that phosphorylates and thereby detoxifies the antibiotic hygromycin B. This gene without a promoter was used to select for endogenous promoter sequences from *C. heterostrophus* that were capable of driving *hygB* expression in the fungus. Transformation frequencies as high as 10 per microgram of DNA were obtained (Turgeon et al., 1987), although frequencies 10 times higher are now possible (O.C. Yoder, personal communication). *C. carbonum* has also been transformed at low frequencies (S.P. Briggs and O.C. Yoder, personal communication). Transformation vectors containing *hygB* and a *Cochliobolus* promoter integrate at the corresponding promoter in the recipient (Turgeon et al., 1987; O.C. Yoder, personal communication).

Complementation has also been used as a selection scheme. Turgeon et al. (1986) cloned a gene involved in tryptophan biosynthesis by complementation of an *E. coli trpF* mutant. This cloned tryptophan gene, called *TRP1*, corresponds functionally to the *trp-1* gene of *N. crassa* and to the *trpC* gene of *A. nidulans* (Turgeon et al., 1986). Using the cloned *TRP1* gene on a plasmid also containing *hygB*, it was shown that homologous recombination occurs in *C. heterostrophus* at high frequency at *TRP1*; therefore, gene disruption and gene replacement are feasible. Homologous double crossover, which yields a true gene replacement event, has been reported (Mullin et al., 1988).

Transforming DNA is stable both mitotically and meiotically in *Cochliobolus*, unlike in *Neurospora* (Selker et al., 1987). Foreign DNA in *C. heterostrophus* is stable through at least five transfers on complete growth medium in the absence of hygromycin (Keller et al., 1988). The foreign DNA is also relatively stable when transformants are passaged five times in the host, maize, although one chromosomal configuration of the integrated DNA was unstable (Keller et al., 1988; O.C. Yoder, personal communication). The biological basis for the differential stability of the *hygB* gene is unknown, nor is it yet known how *hygB* gene stability might relate to the stability of other genes, for example, those involved in pathogenicity. The prevalence of race T over race O (which does not produce T-toxin) declined after T-cms maize was no longer grown commerically (Leonard, 1977a). Race O has a greater fitness than race T on N cytoplasm maize, perhaps because of the metabolic costs associated with producing T-toxin (Blanco and Nelson, 1972; Leonard, 1977b; Klittich and Bronson, 1986).

IV. BIOCHEMISTRY

Work has recently started on the biosynthesis of the *Cochliobolus* host-selective toxins. [^{3}H] leucine is incorporated in vivo into victorin, which contains hydroxyleucine and dichloroleucine (Gloer et al., 1985). The aglycone (a sesquiterpenoid) of HS-toxin (helminthosporoside) has been tritiated, fed to *C. sacchari*, and labeled HS-toxin and homologs containing fewer than four galactose residues have been isolated and characterized (Nakajima and Scheffer, 1987). [5-^{13}C] mevalonic acid, when fed to *H. sacchari*, is incorporated as predicted into the aglycone of HS-toxin (Macko et al., 1983).

Two proteins involved in the biosynthesis of HC-toxin have been identified (Walton, 1987) by following the precedents from work on the biosynthesis of other cyclic peptides from bacteria and fungi (Kleinkauf and von Dohren, 1981, 1987; Zocher et al., 1982). One, called HTS-1, catalyzes an ATP/pyrophosphate (PP$_i$) exchange reaction dependent on L-alanine and D-alanine; the other, HTS-2, catalyzes an ATP/PP$_i$ exchange that is dependent on L-proline. Both activities are present in races and isolates of *C. carbonum* that make HC-toxin and are absent in isolates that do not make HC-toxin. Both proteins have apparent molecular weights of 310,000 by gel filtration (Walton, 1987), but SDS-PAGE indicates that HTS-1 has a molecular weight of 220,000 and HTS-2 a molecular weight of 160,000 (Walton and Holden, 1988). Wessel et al. (1987, 1988a) have also found D-alanine, L-alanine, and L-proline ATP/PP$_i$ exchange activities in *C. carbonum*.

Like other cyclic peptide synthetases, HTS-1 and HTS-2 form thioesters with their amino acid substrates. At some stage of thioesterification of amino acids, HTS-1 converts L-proline to D-proline, the isomer of proline found in HC-toxin (Walton et al., 1982; Walton and Holden, 1988). HTS-2 converts L-alanine to D-alanine, but does not convert D-alanine to L-alanine. These results are consistent with feeding studies using radiolabeled amino acids. [^{14}C] L-proline is incorporated into HC-toxin in vivo reasonably well, but [¹⁴C] D-alanine is incorporated very efficiently (Figure 3A), presumably because it is metabolized poorly if at all. Incorporation of radiolabeled L-alanine in vivo is very poor, only about one one-hundreth that of D-alanine, as expected from its ability to be incorporated into pro-

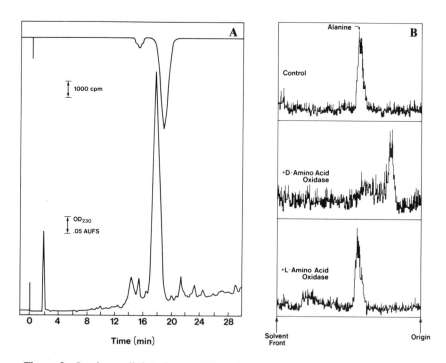

Figure 3 In vivo radiolabeling of HC-toxin. **A:** Reverse-phase (C18) HPLC of HC-toxin radiolabeled with [¹⁴C] D-alanine in vivo after chloroform extraction (Walton et al., 1982). Upper trace (pen polarity is reversed relative to the lower trace): Radioactivity measured with an in-line radioactivity detector. Lower trace: Absorption at 230 nm. The difference in the apparent elution time between the major peaks of UV absorbance and radioactivity is due to chart recorder pen offset and the contained liquid volume between the UV monitor and the radioactivity detector. **B:** Evidence that [¹⁴C] D-alanine labels only the D-alanine residue in HC-toxin. HC-toxin prepared as in **A** was hydrolyzed to its constituent amino acids in 6 N HCl, treated with D-amino acid oxidase or L-amino acid oxidase (Sigma), and analyzed by silica TLC. Radioactivity in each lane was then measured with a TLC plate scanner. The radioactivity in HC-toxin is completely digested by D-amino acid oxidase, as judged by its shift in the migration distance (middle panel), but not at all by L-amino acid oxidase (lower panel).

teins and to be metabolized to pyruvate. Radiolabeled D-alanine is incorporated only into the D-alanine and not into the L-alanine residue of HC-toxin (Figure 3B). Thus, L-alanine is probably the natural precursor to both alanines in HC-toxin. D-alanine could be a natural precursor only to the D-alanine residue, and then only if the fungus had some means other than HTS-2 to obtain it.

A schematic picture of our knowledge of the functional organization of HTS-1 and HTS-2 is given in Figure 4. In the case of HTS-2 (lower figure),

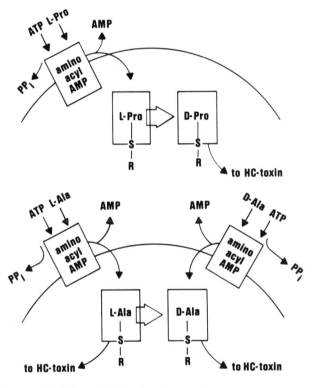

Figure 4 Schematic functional models of HTS-1 (top) and HTS-2 (bottom). HTS-1 and HTS-2 are two proteins with molecular weights by SDS-PAGE of 220,000 and 160,000, respectively (Walton and Holden, 1988). HTS-1 activates L-proline but not D-proline; HTS-2 activates both L-alanine and D-alanine. HTS-1 can epimerize L-proline to D-proline. HTS-2 can epimerize L-alanine to D-alanine, but not D-alanine to L-alanine (Walton and Holden, 1988). It is not known at what step epimerization occurs.

we know that there must be two independent sites for the initial recognition of L-alanine and D-alanine, because in the ATP/PP$_i$ assay the activity in the presence of both amino acids is the sum of the activity in the presence of each amino acid separately. We do not yet know, however, how many thioester sites there are for proline and alanine on HTS-1 and HTS-2, respectively.

HTS-1 and HTS-2 have exceptionally low affinities for their amino acid substrates, much lower than other cyclic peptide synthetases for their amino acids. HTS-1 has a K$_m$ for L-proline of 18 mM, and HTS-2 has K$_m$'s for D-alanine and L-alanine of 3.4 mM and 101 mM, respectively, when measured at a saturating concentration of ATP in the ATP/PP$_i$ exchange assay (Walton and Holden, 1988). We feel that these results are probably real, for several reasons. First, the K$_m$'s for PP$_i$ and ATP are both about 0.2 mM; these are physiologically reasonable and are in agreement with the K$_m$'s of other cyclic peptide synthetases (Kittelberger et al., 1982). Second, plots of enzyme velocity vs. amino acid substrate concentration give normal hyperbolic Michaelis-Menten curves and normal linear double reciprocal plots. This is shown in Figure 5 for HTS-2 and L-alanine. Third, other enzymes are known with K$_m$'s for L-alanine in the high millimolar range (cited in Walton and Holden, 1988). Fourth, it is possible to rationalize the high K$_m$'s of HTS-1 and HTS-2 for their amino acid substrates as a biochemical mechanism to separate secondary from primary metabolism (Walton and Holden, 1988).

The low affinities of HTS-1 and HTS-2 for their amino acid substrates suggested that perhaps supplementing the medium with proline and alanine would stimulate HC-toxin production. We have attempted this without success: Either the exogenous amino acids do not enter the cells or their intracellular concentrations are not limiting to HC-toxin biosynthesis. The concentrations of free proline and alanine in *C. carbonum* have been measured and are unremarkable when compared with either levels of the other amino acids or with levels of alanine and proline in other fungi (Holden, 1962).

Peptide chain formation always begins with phenylalanine during the biosynthesis of gramicidin or tyrocidine (Kleinkeuf and von Döhren, 1987). It is not known in what order the constituent amino acids of HC-toxin are polymerized to form the tetrapeptide. Little is known about the biosynthesis and activation of AOE, the unusual epoxide-containing amino acid in HC-toxin. Radiolabeled acetate is incorporated into AOE in vivo in short-term feeding experiments (Wessel et al., 1988b), suggesting that AOE has a polyketide origin. This possibility is being tested by feeding

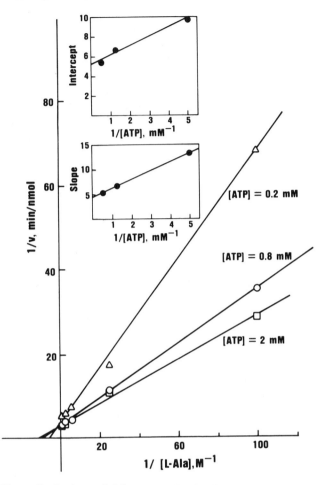

Figure 5 Reciprocal (Lineweaver-Burk) plots of enzyme velocity (ATP/PP$_i$ exchange assay) vs. L-alanine concentration for HTS-2 at three different ATP concentrations.

acetate labeled with ^{13}C to the fungus and then analyzing the labeled HC-toxin by NMR (Pocard, Nair, and Walton, unpublished). We have searched unsuccessfully for enzymatic activities that are characteristic of polyketide synthetases and are unique to race 1 of *C. carbonum*. Among other possible reasons for this failure, AOE biosynthesis might be common to all the isolates of *C. carbonum*.

The discovery that at least two independent proteins are necessary for the biosynthesis of HC-toxin bears on the previous finding that a single genetic locus controls the production of HC-toxin (Scheffer et al., 1967). This locus, called *Tox2* to distinguish it from *Tox1* controlling T-toxin production in *C. heterostrophus*, may be complex. Several examples of gene clustering have been studied in filamentous fungi (Arst and Scazzochio, 1985; Huiet and Case, 1985). Genes for the production of antibiotics are often clustered in *Streptomyces* (e.g., Motamedi and Hutchinson, 1987).

V. STRATEGIES FOR CLONING THE *Tox* LOCI FROM *COCHLIOBOLUS*

If we could clone the *Tox* genes from *Cochliobolus* we would be able to address many interesting questions. For example, why does *Cochliobolus*, among all plant pathogens with the exception of *Alternaria*, specialize in producing host-selective toxins? How and when did species of *Cochliobolus* acquire the ability to produce their particular host-selective toxins? How does this relate to the inadvertent introduction of genes conferring toxin sensitivity into crop plants? Are the different *Tox* loci related evolutionarily?

Relevant to these points is the ability of other fungi to produce analogs of some of the host-selective toxins. There are three analogs of HC-toxin that are produced by three different, unrelated fungi (for structures, see Walton et al., 1985 and Kawai et al., 1986). *Phyllosticta maydis*, another selective pathogen of T-cms maize, makes compounds closely related to T-toxin (Danko et al., 1984). Either the ability of unrelated fungi to make closely related compounds has evolved independently (four times in the case of HC-toxin and its analogs), the ability evolved once in a common ancestor and has been lost from all other fungi, or the ability has been transferred from one fungus to another. Hyphal anastomosis is one means by which filamentous fungi could, in theory, transfer genes between species. Such "lateral" gene transfer (Luckner, 1984), being presumably a rare event, would favor the transfer of traits, such as host-selective toxin production, that are encoded by single or closely linked genes.

Two strategies are currently being employed to clone *Tox* loci from *Cochliobolus*. Yoder and co-workers have made a library of total genomic DNA from a *Tox1+* strain of *C. heterostrophus* in a cosmid vector containing the *hygB* gene and have transferred the entire library into a *Tox1−* strain. The transformants were originally being screened by testing each one for increased virulence on T-cms maize. Recently, however, a strain of *E. coli* expressing the 13,000-Da protein that confers sensitivity to T-toxin on maize mitochondria (Dewey et al., 1988) has proven useful. An agar plug on which each transformant fungus is growing is placed on a lawn of this *E. coli* strain, and toxin production is apparent within 12 hr as a zone of inhibition of bacterial growth (O.C. Yoder, personal communication).

Knowing that the insert size in the cosmid vector is 40,000 bp, assuming that *Tox1* is small relative to the insert size, and estimating that the *C. heterostrophus* genome is 2.6×10^7 bp (similar in size to that of *Aspergillus* and *Neurospora*), then approximately one transformant in every 3100 independent transformants should receive the *Tox1* allele. However, if the *Tox* locus is large, then fewer cosmids will contain the entire locus and more transformants will have to be screened. For example, one would have to screen approximately 6100 independent transformants to find a 20-kb gene (Clarke and Carbon, 1976). In the case of HC-toxin, the combined size of the two enzymes already described would require one gene of about 11 kb (discounting introns), and HC-toxin biosynthesis probably requires other as yet undiscovered enzymes.

The strategy being used in our laboratory to clone the *Tox2* locus from *C. carbonum* depends on identifying and purifying the putative gene products. Antibodies raised against the purified products have been used to screen a cDNA library made in the expression vector lambda gt11. An alternative (but ultimately complementary) approach has been to screen DNA libraries with synthetic oligonucleotide probes based on amino acid sequences from the proteins. The major assumption behind both of these approaches is that the *Tox2* locus encodes biosynthetic enzymes and not a regulatory protein. This approach also depends on the successful purification of unstable proteins that are present in small amounts. Furthermore, it lacks the universal applicability of the transformation approach—it may be impossible to identify and purify enzymes involved in the biosynthesis of T-toxin, victorin, or HS-toxin.

VI. POTENTIAL COMMERCIAL USEFULNESS OF *COCHLIOBOLUS*

Cochliobolus grows quickly, is not nutritionally fastidious, is genetically stable, has a sexual stage for Mendelian genetic analyses, and is transformable for molecular analyses. In these regards it is not unique and probably offers no advantages over other filamentous fungi that are biochemically and genetically better understood. *Cochliobolus* stands out from other fungi, including other pathogens, by its ability to produce host-selective toxins and to exploit the genetic vulnerability of crop plants. Could this be due to an unusually large variety of secondary metabolic pathways, an unusual ability to generate such pathways, or perhaps an ability to acquire novel pathways from other organisms? If they exist, such traits could have commercial utility. For now, however, we still lack sufficient knowledge to be able to conclude in what ways *Cochliobolus* might have something to offer as an industrial fungus.

ACKNOWLEDGMENTS

Work in the author's laboratory on HC-toxin synthetase has been supported by the National Science Foundation (DMB 87-15608) and the Department of Energy Division of Biological Energy Research (Contract DE-ACO2-76ER01338).

REFERENCES

Alcorn, J.L. (1983). Generic concepts in *Drechslera, Bipolaris,* and *Exserohilum. Mycotaxon 17*:1-86.

Aldrich, H.C., Gracen, V.E., York, D., Earle, E.D., and Yoder, O.C. (1977). Ultrastructural effects of *Helminthosporium maydis* race T toxin on mitochondria of corn roots and protoplasts. *Tissue & Cell 9*:167-177.

Arst, H.N., Jr. and Scazzocchio, C. (1985). Formal genetics and molecular biology of the control of gene expression in *Aspergillus nidulans*. In *Gene Manipulations in Fungi*, Bennett, J.W. and Lasure, L.L., eds., Academic Press, Orlando, FL, p. 309-343.

Blanco, M.H. and Nelson, R.R. (1972). Relative survival of populations of race O and race T of *Helminthosporium maydis* on a corn hybrid in normal cytoplasm. *Plant Dis. Rep. 56*:889-891.

Braun, C.J., Siedow, J.N., Williams, M.E., and Levings, C.S. III. (1989). Mutations in the maize mitochondrial T-urf13 eliminate sensitivity to a fungal pathotoxin. *Proc. Natl. Acad. Sci. USA 86*:4435-4439.

Ciuffetti, L.M., Pope, M.R., Dunkle, L.D., Daly, J.M., and Knoche, H.W. (1983). Isolation and structure of an inactive product derived from the host-specific toxin produced by *Helminthosporium carbonum. Biochemistry 22*:3507-3510.

Clarke, L. and Carbon, J. (1976). A colony bank containing synthetic Col El hybrid plasmids representative of the entire *E. coli* genome. *Cell 9*:91-99.

Daly, J.M. (1981). Mechanisms of action. In *Toxins in Plant Disease*, Durbin, R.D., ed., Academic Press, New York, p. 331-394.

Danko, S.J., Kono, Y., Daly, J.M., Suzuki, Y., Takeuchi, S., and McCrery, D.A. (1984). Structure and biological activity of a host-specific toxin produced by the fungal corn pathogen *Phyllosticta maydis. Biochemistry 23*:759-766.

Dewey, R.E., Levings, C.S. III, and Timothy, D.H. (1986). Novel recombinations in the maize mitochondrial genome produce a unique transcriptional unit in the Texas male-sterile cytoplasm. *Cell 44*:438-449.

Dewey, R.E., Siedow, J.N., Timothy, D.H., and Levings, C.S. III (1988). A 13-kilodalton maize mitochondrial protein in *E. coli* confers sensitivity to *Bipolaris maydis* toxin. *Science 239*:293-295.

Dimroth, P., Ringelmann, E., and Lynen, F. (1976). 6-methylsalicylic acid synthetase from *Penicillium patulum*. *Eur. J. Biochem. 68*:591-596.

Duvick, J.P., Daly, J.M., Kratky, Z., Macko, V., Acklin, W., and Arigoni, D. (1984). Biological activity of the isomeric forms of *Helminthosporium sacchari* toxin and of homologs produced in culture. *Plant Physiol. 74*:117-122.

Earle, E.D. and Gracen, V.E. (1982). Effects of *Helminthosporium* phytotoxins on cereal leaf protoplasts. In *Proceedings of the 5th International Congress on Plant Tissue and Cell Culture*, p. 663-664.

Elovson, J. (1975). Purification and properties of the fatty acid synthetase complex from *Neurospora crassa*, and the nature of the fas⁻ mutation. *J. Bacteriol. 124*:524-533.

Forde, B.G., Oliver, R.J.C., and Leaver, C.J. (1978). Variation in mitochondrial translation products associated with male-sterile cytoplasms of maize. *Proc. Natl. Acad. Sci. USA 75*:3841-3845.

Forde, B.G. and Leaver, C.J. (1980). Nuclear and cytoplasmic genes controlling synthesis of variant mitochondrial polypeptides in male-sterile maize. *Proc. Natl. Acad. Sci. USA 77*:418-422.

Friedrich, J. (1977). Zur kenntnis der 6-methylsalicylsaure-synthetase aus *Penicillium patulum*. Inaugural-Dissertation, Ludwig-Maximilian· Universität, München, Federal Republic of Germany.

Garber, R.C. and Yoder, O.C. (1983). Isolation of DNA from filamentous fungi and separation into nuclear, mitochondrial, ribosomal, and plasmid components. *Anal. Biochem. 135*:416-422.

Garber, R.C. and Yoder, O.C. (1984). Mitochondrial DNA of the filamentous ascomycete *Cochliobolus heterostrophus*. *Curr. Genet. 8*:621-628.

Garber, R.C., Turgeon, B.G., and Yoder, O.C. (1984). A mitochondrial plasmid from the plant pathogenic fungus *Cochliobolus heterostrophus*. *Mol. Gen. Genet. 196*:301-310.

Gengenbach, B.G., Connelly, J.A., Pring, D.R., and Conde, M.F. (1981). Mitochondrial DNA variation in maize plants regenerated during tissue culture selection. *Theor. Appl. Genet. 59*:161-167.

Gloer, J.B., Meinwald, J., Walton, J.D., and Earle, E.D. (1985). Studies on the fungal phytotoxin victorin: structures of three novel amino acids from the acid hydrolyzate. *Experientia 41*:1370-1374.

Guzman, D., Garber, R.C., and Yoder, O.C. (1982). Cytology of meiosis I and chromosome number of *Cochliobolus heterostrophus* (Ascomycete). *Can. J. Bot. 60*:1138-1141.

Hesseltine, C.W., Ellis, J.J., and Shotwell, O.L. (1971). *Helminthosporium*: secondary metabolites, southern corn leaf blight of corn, and biology. *J. Agr. Food Chem. 19*:707-717.

Holden, J.T., ed. (1962). *Amino Acid Pools*, Elsevier, Amsterdam.

Huiet, L. and Case, M. (1985). Molecular biology of the *qa* gene cluster of Neurospora. In: *Gene Manipulations in Fungi*, Bennett, J.W., and Lasure, L.L., eds., Academic Press, Orlando, FL, p. 229-244.

Kawai, M. and Rich, D.H. (1983). Total synthesis of the cyclic tetrapeptide, HC-toxin. *Tet. Lett. 24*:5309-5312.

Kawai, M., Pottorf, R.S., and Rich, D.H. (1986). Structure and solution conformation of the cytostatic cyclic tetrapeptide WF-3161, cyclo- [L-leucyl-L-pipecolyl-L-(2-amino-8-oxo-9,10-epoxidecanoyl)-D-phenylalanyl]. *J. Am. Chem. Soc. 29*:2409-2411.

Keller, N.P., Bergstrom, G.C., and Yoder, O.C. (1988). Stability of foreign DNA in the *Cochliobolus heterostrophus* genome (abstract). *J. Cell. Biochem. 12C*:282.

Kim, S.-D., Knoche, H.W., Dunkle, L.D., McCrery, D.A., and Tomer, K.B. (1985). Structure of an amino acid analog of the host-specific toxin from *Helminthosporium carbonum*. *Tet. Lett. 26*:969-972.

Kim, S.-D., Knoche, H.W., and Dunkle, L.D. (1987). Essentiality of the ketone function for toxicity of the host-selective toxin produced by *Helminthosporium carbonum*. *Physiol. Mol. Plant Pathol. 30*:433-440.

Kittelberger, R., Altmann, M., and von Dohren, H. (1982). Kinetics of amino acid activation in gramicidin S synthesis. In: *Peptide Antibiotics, Biosynthesis and Functions*, Kleinkauf, H., and von Döhren, H., eds., Walter de Gruyter, Berlin, p. 209-218.

Kleinkauf, H. and von Döhren, H. (1981). Nucleic acid independent synthesis of peptides. *Curr. Top. Microbiol. Immunol. 91*:129-177.

Kleinkauf, H. and von Döhren, H. (1987). Biosynthesis of peptide antibiotics. *Ann. Rev. Microbiol. 41*:259-289.

Kline, D.M. and Nelson, R.R. (1969). Inheritance of factors in *Cochliobolus carbonum* conditioning symptom expression on grass hosts. *Phytopathology 59*: 1133-1135.

Klittich, C.J.R. and Bronson, C.R. (1986). Reduced fitnes associated with *TOX1* of *Cochliobolus heterostrophus*. *Phytopathology 76*:1294-1298.

Kono, Y. and Daly, J.M. (1979). Characterization of the host-specific pathotoxin produced by *Helminthosporium maydis*, race T, affecting corn with Texas male sterile cytoplasm. *Bioorg. Chem. 8*:391-397.

Kono, Y., Kinoshita, T., Takeuchi, S., and Daly, J.M. (1986). Structure of HV-toxin M, a host-specific toxin-related compound produced by *Helminthosporium victoriae*. *Agric. Biol. Chem. 50*:2689-2691.

Kraus, G.A., Silveira, M., and Danko, S.J. (1984). Synthesis and testing of analogues of *Helminthosporium maydis* race T toxin. *J. Agric. Food Chem. 32*: 1265-1268.

Larkin, P.J. and Scowcroft, W.R. (1981). Eyespot disease of sugarcane. *Plant Physiol. 67*:408-414.

Leach, J., Lang, B.R., and Yoder, O.C. (1982a). Methods for selection of mutants and in vitro culture of *Cochliobolus heterostrophus*. *J. Gen. Microbiol. 128*:1719-1729.

Leach, J., Tegtmeier, K.J., Daly, J.M., and Yoder, O.C. (1982b). Dominace at the Tox1 locus controlling T-toxin production by *Cochliobolus heterostrophus. Physiol. Plant Pathol. 21*:327-333.

Leonard, K.J. (1977a). Races of *Bipolaris maydis* in the southeastern U.S. from 1974-1976. *Plant Dis. Rep. 61*:914-915.

Leonard, K.J. (1977b). Virulence, temperature optima, and competitive abilities of isolines of races T and O of *Bipolaris maydis. Phytopathology 67*:1273-1279.

Liesch, J.M., Sweeley, C.C., Staffeld, G.D., Anderson, M.S., Weber, D.J., and Scheffer, R.P. (1982). Structure of HC-toxin, a cyclic tetrapeptide from *Helminthosporium carbonum. Tetrahedron. 38*:45-48.

Lim, S.M. and Hooker, A.L. (1971). Southern corn leaf blight: genetic control of pathogenicity and toxin production in race T and race O of *Cochliobolus heterostrophus. Genetics 69*:115-117.

Lin, W. and Kauer, J.C. (1985). Peptide alcohols as promoters of nitrate and ammonium ion uptake in plants. *Plant Physiol. 77*:403-406.

Livingston, R.S. and Scheffer, R.P. (1981). Isolation and characterization of host-selective toxin from *Helminthosporium sacchari. J. Biol. Chem. 256*:1705-1710.

Livingston, R.S. and Scheffer, R.P. (1983). Conversion of *Helminthosporium sacchari* toxin to toxoids by β-galactofuranosidase from *Helminthosporium. Plant Physiol. 72*:530-534.

Livingston, R.S. and Scheffer, R.P. (1984a). Selective toxins and analogs produced by *Helminthosporium sacchari. Plant Physiol. 76*:96-102.

Livingston, R.S. and Scheffer, R.P. (1984b). Toxic and protective effects of analogues of *Helminthosporium sacchari* toxin on sugarcane tissues. *Physiol. Plant Pathol. 24*:133-142.

Luckner, M. (1984). *Secondary Metabolism in Microorganisms, Plants, and Animals.* Springer, Berlin.

Luttrel, E.S. (1958). The perfect stage of *Helminthosporium turcicum. Phytopathology 48*:281-287.

Macko, V. (1983). Structural aspects of toxins. In: *Toxins and Plant Pathogenesis*, Daly, J.M., and Deverall, B. J., eds., Academic Press, Sydney, p. 41-80.

Macko, V., Goodfriend, K., Wachs, T., Renwick, J.A.A., Acklin, W., and Arigoni, D. (1981). Characterization of the host-specific toxins produced by *Helminthosporium sacchari*, the causal organism of eyespot disease of sugarcane. *Experientia 37*:923-924.

Macko, V., Acklin, W., Hildenbrand, C., Weibel, F., and Arigoni, D. (1983). Structure of three isomeric host-specific toxins from *Helminthosporium sacchari. Experientia 39*:343-347.

Matthews, D.E., Gregory, P. and Gracen, V.E. (1979). *Helminthosporium maydis* race T toxin induces leakage of NAD^+ from T cytoplasm corn mitochondria. *Plant Physiol. 63*:1149-1153.

Miller, R.J. and Koeppe, D.E. (1971). Southern corn leaf blight: susceptible and resistant mitochondria. *Science 173*:67-69.

Motamedi, H. and Hutchinson, C.R. (1987). Cloning and heterologous expression of a gene cluster for the biosynthesis of tetracenomycin C, the anthracycline antitumor antibiotic of *Streptomyces glaucescens*. *Proc. Natl. Acad. Sci. USA 84*:4445-4449.

Mullin, P.G., Turgeon, B.G., Garber, R.C., and Yoder, O.C. (1988). Integration of transforming DNA by homologous recombination allows gene replacement in *Cochliobolus heterostrophus* (abstract). *J. Cell. Biochem. 12C*:285.

Nakajima, H. and Scheffer, R.P. (1987). Interconversions of aglycone and host-selective toxin from *Helminthosporium sacchari*. *Phytochemistry 26*:1607-1611.

Nelson, R.R. and Ullstrup, A.J. (1961). The inheritance of pathogenicity in *Cochliobolus carbonum*. *Phytopathology 51*:1-2.

Nelson, R.R. and Kline, D.M. (1963). Gene systems for pathogenicity and pathogenic potentials. I. Interspecific hybrids of *Cochliobolus carbonum* × *Cochliobolus victoriae*. *Phytopathology 53*:101-105.

Nelson, R.R. and Sherwood, R.T. (1968). Genetic control of endopolygalacturonase production in *Cochliobolus carbonum*. *Phytopathology 58*:1277-1280.

Panopoulos, N.J., Walton, J.D., and Willis, D.K. (1984). Genetic and biochemical basis of virulence in plant pathogens. In: *Genes Involved in Microbe-Plant Interactions*, Verma, D.P.S., and Hohn, T., eds., Springer, Vienna, p. 339-374.

Pope, M.R., Ciuffetti, L.M., Knoche, H.W., McCrery, D., Daly, J.M., and Dunkle, L.D. (1983). Structure of the host-specific toxin produced by *Helminthosporium carbonum*. *Biochemistry 22*:3502-3506.

Rasmussen, J.B. and Scheffer, R.P. (1988). Isolation and biological activities of four selective toxins from *Helminthosporium carbonum*. *Plant Physiol. 86*: 187-191.

Rines, H.W. and Luke, H.H. (1985). Selection and regeneration of toxin-insensitive plants from tissue cultures of oats (*Avena sativa*) susceptible to *Helminthosporium victoriae*. *Theor. Appl. Genet. 71*:16-21.

Rottmann, W.H., Brears, T., Hodge, T.P., and Lonsdale, D.M. (1987). A mitochondrial gene is lost via homologous recombination during reversion of CMS-T maize to fertility. *EMBO J. 6*:1541-1546.

Scheffer, R.P. (1983). Toxins as chemical determinants of plant disease. In: *Toxins and Plant Pathogenesis*, Daly, J.M., and Deverall, B.J., eds., Academic Press, Sydney, p. 1-40.

Scheffer, R.P. and Ullstrup, A.J. (1965). A host-specific toxic metabolite from *Helminthosporium carbonum*. *Phytopathology 55*:1037-1038.

Scheffer, R.P., Nelson, R.R., and Ullstrup, A.J. (1967). Inheritance of toxin production and pathogenicity in *Cochliobolus carbonum* and *Cochliobolus victoriae*. *Phytopathology 57*:1288-1291.

Scheffer, R.P. and Briggs, S.P. (1981). A perspective of toxin studies in plant pathology. In: *Toxins in Plant Disease*, Durbin, R.D., ed., Academic Press, New York, p. 1-20.

Scheffer, R.P. and Livingston, R.S. (1984). Host-selective toxins and their role in plant disease. *Science 223*:17-21.

Schweizer, M., Roberts, L.M., Holtke, J., Takabayashi, K., Muller, G., Hoja, U., Hollerer, E., Schuh, B., and Schweizer, E. (1986). Molecular structure and expression of fatty acid synthetase in yeast. *Biochem. Soc. Trans.* *14*:572-574.

Selker, E.U., Cambareri, E.B., Jensen, B.C., and Haack, K.R. (1987). Rearrangement of duplicated DNA in specialized cells of *Neurospora*. *Cell 51*:741-752.

Sugawara, F., Strobel, G., Strange, R.N., Siedow, J.N., Van Duyne, G.D., and Clardy, J. (1987). Phytotoxins from the pathogenic fungi *Drechslera maydis* and *Drechslera sorghicola. Proc. Natl. Sci. USA 84*:3081-3085.

Suzuki, Y., Knoche, H.W., and Daly, J.M. (1982a). Analogs of host-specific phytotoxin produced by *Helminthosporium maydis*, race T. I. Synthesis. *Bioorg. Chem. 11*:300-312.

Suzuki, Y., Tegtmeier, K.J., Daly, J.M., and Knoche, H.W. (1982b). Analogs of host-specific phytotoxin produced by *Helminthosporium maydis*, race T. II. Biological Activities. *Bioorg. Chem. 11*:313-321.

Tegtmeier, K.J., Daly, J.M., and Yoder, O.C. (1982). T-toxin production by near-iosgenic isolates of *Cochliobolus heterostrophus* races T and O. *Phytopathology 72*:1492-1495.

Tinline, R.D. (1962). *Cochliobolus sativus* V. Heterokaryosis and parasexuality. *Can. J. Bot. 40*:425-437.

Tomas, A. and Bockus, W.W. (1987). Cultivar-specific toxicity of culture filtrates of *Pyrenophora tritici-repentis. Phytopathology 77*:1337-1340.

Turgeon, B.G., Garber, R.C., and Yoder, O.C. (1985). Transformation of the fungal maize pathogen *Cochliobolus heterostrophus* using the *Aspergillus nidulans amdS* gene. *Mol. Gen. Genet. 201*:450-453.

Turgeon, B.G., MacRae, W.D., Garber, R.C., Fink, G.R., and Yoder, O.C. (1986). A cloned tryptophan-synthesis gene from the Ascomycete *Cochliobolus heterostrophus* functions in *Escherichia coli*, yeast and *Aspergillus nidulans. Gene 42*:79-88.

Turgeon, B.G., Garber, R.C., and Yoder, O.C. (1987). Development of a fungal transformation system based on selection of sequences with promoter activity. *Mol. Cell. Biol. 7*:3297-3305.

Turgeon, G., Ciuffetti, L., Schafer, W., and Yoder, O. (1988). Molecular analysis of the mating locus of *Cochliobolus heterostrophus* (abstract). *J. Cell. Biochem. 12C*:292.

Ullstrup, A.J. and Brunson, A.M. (1947). Linkage relationships of a gene in corn determining susceptibility to a *Helminthosporium* leaf spot. *J. Am. Soc. Agron. 39*:606-609.

Vidhyasekaran, P., Borromeo, E.S., and Mew, T.W. (1986). Host-specific toxin production by *Helminthosporium oryzae. Phytopathology 76*:261-266.

Walton, J.D. (1987). Two enzymes involved in biosynthesis of the host-selective phytotoxin HC-toxin. *Proc. Natl. Acad. Sci. USA 84*:8444-8447.

Walton, J.D., Earle, E.D., Yoder, O.C., and Spanswick, R.M. (1979). Reduction of adenosine triphosphate levels in susceptible maize mesophyll protoplasts by *Helminthosporium maydis* race T toxin. *Plant Physiol. 63*:806-810.

Walton, J.D., Earle, E.D., and Gibson, B.W. (1982). Purification and structure of the host-specific toxin from *Helminthosporium carbonum* race 1. *Biochem. Biophys. Res. Comm. 107*:785-794.

Walton, J.D. and Earle, E.D. (1983). The epoxide in HC-toxin is required for activity against susceptible maize. *Physiol. Plant Pathol. 22*:371-376.

Walton, J.D. and Earle, E.D. (1984). Characterization of the host-specific phytotoxin victorin by high-pressure liquid chromatography. *Pl. Sci. Lett. 34*:231-238.

Walton, J.D. and Earle, E.D. (1985). Stimulation of extracellular polysaccharide synthesis in oat protoplasts by the host-specific phytotoxin victorin. *Planta 165*:407-415.

Walton, J.D., Earle, E.D., Stahelin, H., Grieder, A., Hirota, A., and Suzuki, A. (1985). Reciprocal biological activities of the cyclic tetrapeptides chlamydocin and HC-toxin. *Experientia 41*:348-350.

Walton, J.D. and Holden, F.R. (1988). Properties of two enzymes involved in the biosynthesis of the fungal pathogenicity factor HC-toxin. *Mol. Plant-Microbe Interactions 1*:128-134.

Wessel, W.L., Clare, K.A., and Gibbons, W.A. (1987). Enzymatic biosynthesis of peptide toxins by plant-specific fungi. *Biochem. Soc. Trans. 15*:917-918.

Wessel, W. L., Clare, K.A., and Gibbons, W.A. (1988a). Purification and characterization of HC-toxin synthetase from *Helminthosporium carbonum*. *Biochem. Soc. Trans. 16*:401-402.

Wessel, W.L., Clare, K.A., and Gibbons, W.A. (1988b). Biosynthesis of L-AEO, the toxic determinant of the phytotoxin produced by *Helminthosporium carbonum*. *Bichem. Soc. Trans. 16*:402-403.

Wise, R.P., Pring, D.R., and Gengenbach, B.G. (1987). Mutation to male fertility and toxin insensitivity in Texas (T)-cytoplasm is associated with a frameshift in a mitochondrial open reading frame. *Proc. Natl. Acad. Sci. USA 84*:2858-2862.

Wolpert, T.J., Macko, V., Acklin, W., Juan, B., Seibl, J., Meili, J., and Arigoni, D. (1985). Structure of victorin C, the major host-selective toxin from *Cochliobolus victoriae*. *Experientia 41*:1524-1529.

Wolpert, T.J., Macko, V., Acklin, W., Juan, B., and Arigoni, D. (1986). Structure of minor host-selective toxins from *Cochliobolus victoriae*. *Experientia 42*:1296-1299.

Wolpert, T.J. and Macko, V. (1989). Specific binding of victorin to a 100-kDa protein from oats. *Proc. Natl. Acad. Sci. USA 86*:4092-4096.

Yoder, O.C. (1980). Toxins in pathogenesis. *Ann. Rev. Phytopathol. 18*:103-129.

Yoder, O.C. (1988). *Cochliobolus heterostrophus*: cause of Southern corn leaf blight. *Adv. Plant Pathol. 6*:93-112.

Yoder, O.C. and Scheffer, R.P. (1973a). Effects of *Helminthosporium carbonum* toxin on nitrate uptake and reduction by corn tissues. *Plant Physiol. 52*:513-517.

Yoder, O.C. and Scheffer, R.P. (1973b). Effects of *Helminthosporium carbonum* toxin on absorption of solutes by corn roots. *Plant Physiol. 52*:518-523.

Yoder, O.C. and Gracen, V.E. (1975). Segregation of pathogenicity types and host-specific toxin production in progenies of crosses between race T and O of *Helminthosporium maydis* (*Cochliobolus heterostrophus*). *Phytopathology* *65*:273-276.

Yoder, O.C., Valent, B., and Chumley, F. (1986). Genetic nomenclature and practice for plant pathogenic fungi. *Phytopathology 76*:383-385.

Zocher, R., Keller, U., and Kleinkauf, H. (1982). Enniatin synthetase, a novel type of multifunctional enzyme catalyzing depsipeptide synthesis in *Fusarium oxysporum*. *Biochemistry 21*:43-48.

10

Scaling Up Production of Recombinant DNA Products Using Filamentous Fungi as Hosts

Nigel J. Smart*
Allelix Biopharmaceuticals, Mississauga, Ontario, Canada

I. INTRODUCTION

Traditionally, filamentous fungi have been used as producers of antibiotics, bulk enzymes, and specialty chemicals, examples of which are indicated in Table 1. More recently, due to advances in the molecular biology of filamentous fungi, these systems are now being considered as suitable alternatives to *Escherichia coli* and *Saccharomyces cerevisiae* for the industrial production of recombinant proteins. Their value as production hosts for modern biotechnology applications is derived from the experience gained from a well-cataloged biological history, and some of these factors are listed in Table 2. In terms of their use as producers of heterologous proteins, several criteria are important:

1. It is possible to generate a high expression level of the recombinant protein.
2. In vivo modification of the gene products is possible to produce molecules with similar physical and biological properties as the native molecules. Of the factors listed in Table 2, secretion (and secretory mecha-

---

Current affiliation: SmithKline Beecham Pharmaceuticals, King of Prussia, Pennsylvania

Table 1   Examples of Products Traditionally Made
Using Filamentous Fungi

Product	Species
Penicillin G	*Penicillium chrysogenum*
Cephalosporin C	*Acremonium chrysogenum*
Griseofulvin	*Penicillium patulum*
Citric acid	*Aspergillus niger*
Gluconic acid	*Aspergillus niger*
Itaconic acid	*Aspergillus terreus*
Proteases	*Aspergillus oryzae*
Amyloglucosidase	*Aspergillus niger*
Alkaloids	*Claviceps purpurea*
Steroid hydroxylations	*Coniothyrium hellbori*

Table 2   Advantages of the *Aspergilli* as Expression/Secretion Hosts

• Eukaryotic host
• Good protein secretor
• F.D.A. approved (*A. niger*)
• Accepted by industry (organic acid and industrial enzyme production)
• Rapid growth on simple, inexpensive media
• Well characterized genetically (*A. nidulans*)
  — available promoters with well-defined regulation
  — many mutants
• Mitotically stable integrated transformants

nisms) is presently an issue that is receiving considerable attention, since
it avoids several problems (identified in Table 3) that make the production of heterologous proteins as much an art as a science.

In terms of bioprocessing, purification should be easier due to the use of
simple defined media and the avoidance of disruption procedures, which
often cause structural proteins to be released in the case of intracellular
production. Other advantages are listed in Table 4.

Process scale-up of microbial systems is an extremely complicated blend
of science and engineering, requiring a knowledge base that extends from
physiology to chemical engineering. Furthermore, exploitation of recombinant systems adds an additional dimension, since this demands an understanding of the molecular biology of the expression system and the protein
biochemistry of the gene product. The overriding factor that makes scale-

Table 3 Biological Advantages of Extracellular Secretion as a Method of Protein Production

- Avoids intracellular accumulation of toxic levels of product
- Potentially avoids intracellular hydrolysis or undesirable modification of product
- Takes advantage of posttranslational events
  - glycosylation
  - protein folding
  - other modifications
- Potential use of directed endoproteolysis

Table 4 Technical Advantages of Extracellular Secretion as a Method of Protein Production

- Ease of purification (affinity chromatography of media, etc.)
- Continuous fermentation is possible
- Avoids crystal formation, resulting in inactive, insoluble product

up difficult to predict and that often results in a rather empirical solution, is that perfect macromixing is impossible to achieve because concentration gradients develop in translation from the laboratory scale to the production scale.

In selecting filamentous fungi as hosts for heterologous proteins, several key issues need to be addressed, relating to the technical feasibility of using these organisms as expression hosts and regulatory issues that may become significant, especially in the case of pharmaceutical proteins.

Technical issues include:

1. the rheology of the biological fluid
2. the overall hydrodynamics of the bioreactor system
3. the oxygen transfer of the bioreactor system

In the case of the production of recombinant proteins, many other factors affect fermentation yield, and some of these are listed in Table 5. However, in discussing the development of these systems for recombinant production, it is worthwhile to first review some of the general principles that affect these systems before highlighting details more specific to genetically engineered systems. For this purpose, examples will be drawn from our own experience using Allelix's proprietary *Aspergillus* expression system.

Table 5 Factors Affecting Product
Expression

1. Culture pH
2. Gas transfer — direct and indirect
3. Agitation — shear
4. Metal ions — filament branching
5. Carbon source selection
6. Inducer selection
7. Inducer manipulation
8. Postinduction temperature

II. TECHNICAL ISSUES RELATING TO SCALE-UP

A. Rheology of Biological Fluids

Like all fermentation processes involving filamentous fungi, the culture
fluid of these organisms is very viscous and exhibits behavior often referred
to as "non-Newtonian" in nature.

In classical terms, these fluids do not possess a true viscosity but instead exhibit an apparent viscosity that is dependent upon the shear rate
applied. This is not a constant characteristic property of non-Newtonian
fluids, as illustrated in Figure 1, where shear stress is plotted against shear
rate on Cartesian coordinates (Blanch and Bhavaraju, 1976). The relationship depicted is known as a *rheogram*, since it characterizes the "rheo-
logical properties" of the fluid. As a result of this property, the normal
processes of heat and mass transfer becomes more difficult as the concentration of cells in the bioreactor increase, and this increases in magnitude as the scale of operation is increased.

In general, Newtonian fluids may be described in terms of their rheo-
logical behavior by equation (1), defined by Margaritis and Zajic (1978):

$$\tau = \tau_1 + K(D)^{n*} \tag{1}$$

where τ = shear stress (mPa·s)
   τ_1 = yield stress (PaNm^{-2})
   k = power law constant
   $n*$ = flow behavior
   D = shear rate (s^{-1})

In this equation, τ_1, k, and $n*$ represent specific constants for a given
fluid, and for Newtonian behavior, $\tau_1 = 0$ and $n* = 1$.

However, for fungal fermentation fluids that exhibit non-Newtonian
behavior, a series of modifications exists that depends on the particular

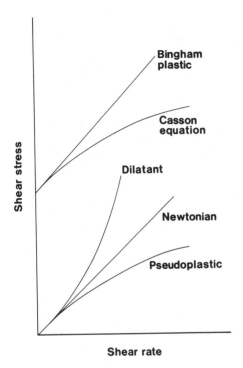

Figure 1 Flow behavior of fluids.

subgrouping. In the case of "power law fluids," where $\tau_1 = 0$ and $n < 1$, these are referred to as *pseudoplastic fluids*; whereas in cases where $\tau_1 = 0$ and $n > 1$, these are termed *dilatant fluids*. In cases where $\tau_1 = 0$ and $n = 1$, however, these are termed *Bingham plastics*.

Fluids that are characterized as Casson fluids require a slightly modified expression and this is defined in Eq. (2).

$$\tau = \tau_1 + K_c (D) \qquad (2)$$

where K_c = Casson constant

On a macroscopic scale, the branched mycelial network forms a three-dimensional structure that imparts rigidity to the suspension. As such, these fluids tend to exhibit a well-defined yield stress.

Furthermore, the relatively long hyphae seem to align themselves with each other at high shear rates, with the result that the apparent viscosity decreases with increasing shear rate and a form of thixotropic-type flow behavior is produced. This type of behavior is often observed when the agitator tip speed is increased, with the result that they tend to be characterized as either Bingham plastics, pseudoplastic, or Casson-type fluids.

During the time course of a fermentation, the rheology changes due to several factors, which include an increase in the cell density, as noted by the mycelial weight and changing morphological characteristics, as quantified by the degree of branching.

Usually, the early stages of a fermentation are characterized by the spores germinating, then being dispersed and forming tiny aggregates. While the organism is in this early stage of development, the rheology is Newtonian. However, as the culture grows and a three-dimensional branched mycelial network is formed, the culture transforms into a pronounced non-Newtonian form. Later on, as the culture starts to lyse and the branching becomes fragmented, the culture reverts to a form of semi-Newtonian behavior again. In the case of recombinant cultures of *Aspergillus*, this is often during the postinduction phase when the cells are in a semistarved nutritional state.

B. Hydrodynamics of Bioreactor Systems

This changing rheology also has a marked effect on the amount of power that needs to be supplied to the culture to maintain agitation, and this can also have an impact on the degree of branching if the shear rate is maintained at a high level. In general, as the liquid volume increases with scale-up, the power input for agitation cannot be maintained in relative proportion, due to the economic and physical limitations connected with the sizing of the agitation drive train, in addition to an increasing relative sensitivity of some of the cells to hydrodynamic shear effects. This frequently manifests itself in the form of broken mycelial filaments. This is illustrated in Figure 2A and 2B.

In practice, as the cells circulate in the vessel bulk fluid, their local "fluid element" environment is enriched as they move within the vicinity of the impeller system. Then, as they move away from the impeller, they move into an unmixed region. However, this also causes an oscillation in the degree of shear stress on the cells, and this can cause oscillations in cellular metabolism, because of the interaction between the well-mixed and unmixed regions (Einsele et al., 1978). In many cases, the effects of this fluctuating shear force is uncharacterized, but it usually results in reduced overall productivity.

Although maximum shear rate is a critical parameter because it influences filament size, on scale-up the average shear rate often becomes a more important factor, because the residence time of any given culture element is reduced in relative terms compared to the case for a small lab-

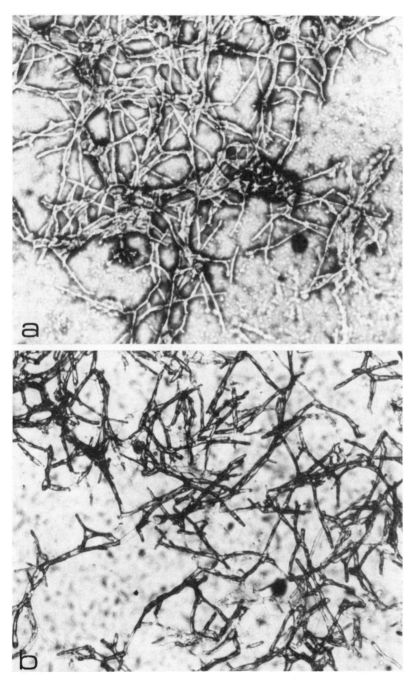

Figure 2 Intact and shear damaged filaments of a recombinant *A. nidulans*. (a) Intact filaments; (b) broken filaments; 300 magnification.

oratory fermenter. Hence, the average shear rate is often the parameter that influences cellular morphology.

Average shear rate, which is a function of impeller speed, is defined by the following equation (Metzner and Otto, 1957):

$$\dot{\gamma} = KN \tag{3}$$

where N = impeller rotational speed
K = proportionality constant
$\dot{\gamma}$ = average shear rate

Normally, standard practice in the field of fermentation technology is to endeavor to work in the region of turbulent flow. However, mold fermentation often prevents these conditions from being fully established. Instead the flow conditions usually lie in a transition region between viscous and turbulent flow. From their studies using pseudoplastic studies with flat turbine impellers, Metzner and Otto (1957) found the proportionality factor that relates average shear and impeller shaft speed to have a value of approximately 13.

C. Oxygen Transfer in Bioreactor Systems

Dispersion and the transmission of power into a large-scale fungal fermentation system is extremely important, since this determines the rate of growth of the culture, its viability, and its capability to synthesize the product. In general, cells starved of nutritionally important oxygen are incapable of producing gene products in appreciable amounts (Smart, 1988, Allelix unpublished data).

In submerged fermentation fluids, the oxygen supplied to the culture must first dissolve before it is available for use by the cells. This rate of solution is governed by Eq. (4), which describes the relationship between the oxygen solution rate and the volumetric transfer coefficient – $K_L a$.

$$\text{OSR} = K_L a(C^* - C) \tag{4}$$

where OSR = oxygen solution rate
K_L = mass transfer coefficient
a = gas/liquid interfacial area
C^* = concentration of dissolved oxygen at equilibrium
C = concentration of dissolved oxygen in the bulk of the liquid

This volumetric transfer coefficient is universally employed as a measure of aeration (ventilation) efficiency, since it is independent of the dissolved oxygen concentration, and it has dimensions of reciprocal time $-h^{-1}$.

In Eq. (4), the difference in oxygen concentrations represents the driving force across this resistance, and in highly viscous fungal fermentations, where oxygen may be limited in the bulk of the culture fluid, this equation may be reduced to produce a simpler relationship, as illustrated in Eq. (5).

$$OSR = K_L aC^* \tag{5}$$

Many factors are known to influence aeration (ventilation), including agitation, air flow rate, pressure, temperature, bioreactor configuration, fluid type, media constituents, and defoaming agents. However, by far the most important is agitation. It is therefore important to produce a regime that is balanced in terms of supplying adequate agitation to prevent microenvironment formation, but insufficient to cause filament damage.

In practice, this is quite a complicated task, since the morphology and physiology of culture is dynamic, with the result that there is a changing requirement for oxygen and for bulk mixing. Frequently these are not in phase with each other, and this can be illustrated when the culture viscosity is still high and requires a substantial power input to maintain adequate bulk mixing, but the oxygen demand of the cells has diminished. This is illustrated schematically in Figure 3.

A similar and related problem occurs in relation to meeting the oxygen demand as the culture grows. This occurs as a result of the non-Newtonian

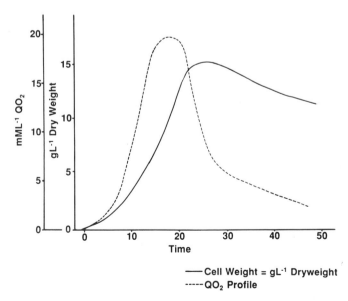

Figure 3 Schematic illustration of the decreased respiratory demand for oxygen for a culture of recombinant *Aspergillus nidulans* at high cell density.

characteristics, which cause the rate of oxygen gas transfer to fall as the fluid viscosity increases. Frequently this results in oxygen-starved cultures on scale-up, even though techniques such as tank overpressurization are used to increase the driving force to gaseous desolution. This decrease in gas transfer has been noted by Deindoerfer and Gaden (1955) and is illustrated in Figure 4.

III. SPECIFIC PROBLEMS OF RECOMBINANT FUNGI

In relation to specific details connected with the expression and scale-up of recombinant fungal systems, some of these will now be illustrated with reference to the Allelix *Aspergillus* system.

A. The Expression System

In brief, the system is based upon DNA cartridges that contain the alcohol dehydrogenase (*alcA*) promoter and a nucleotide sequence that encodes a fungal secretion signal (Gwynne et al., 1987). More specifically, this cartridge contains a promoter from the alcohol dehydrogenase I locus of *Aspergillus nidulans*, the 5′ untranslated sequence from the natural alcohol

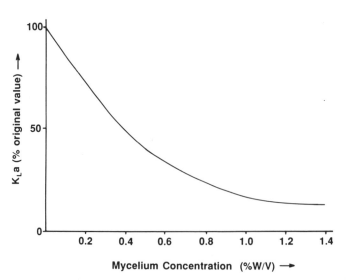

Figure 4 Decrease in the volumetric oxygen mass transfer coefficient with increasing "apparent" fluid viscosity.

dehydrogenase I gene, a sequence encoding a secretion signal that functions in *Aspergillus n.*, and a sequence from the 3' untranscribed region of the *Aspergillus n.* glucoamylase gene, which is used as a transcriptional terminator. The gene of interest, fused with the cartridge to the secretion signal sequence, is prepared in a bacterial plasmid and is inserted into suitable host cells of *Aspergillus* by a process of cotransformation, with a second plasmid containing a selectable marker. Transformants are then identified using standard techniques for the presence of the integrated expression assembly. Duplication of the "new" DNA occurs thereafter as part of the host's chromosomal DNA, which consequently ensures its stability.

AlcA expression is tightly regulated by the *alcR* gene, as illustrated in Figure 5, Gwynne, 1987. The *alcR* gene encodes a *trans*-acting factor (or positive effector) that is absolutely required for *alcA* activity. In practice it has been found that, even with high *alcA* copy number, product secretion is low unless there are multicopies of the *alcR* gene. This is illustrated in Figure 6. One explanation offered to account for this is that there is some titrating out of the *trans*-acting factor occurring at a low *alaR* copy number. In practice, the *Aspergillus* transformants are usually grown to high cell density in a simple media before being switched on by an *alcA* inducer to initiate product expression. This inducer is often a simple industrial ketone.

B. Process Kinetics

The efficiency of the expression system can be illustrated by referring to fermentation data from three separate gene constructs, producing products with a range of molecular weights. For the purpose of being illustrative, these molecules have molecular weights of approximately 67, 6, and

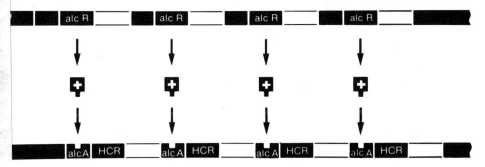

Figure 5 Multicopy activation of the fungal *alaA* promoter.

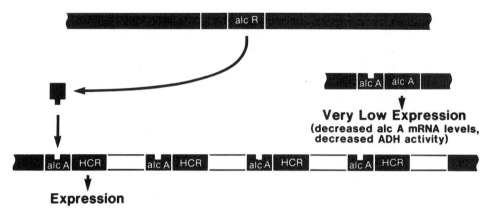

Figure 6 Single-copy activation of the fungal *alaA* promoter.

16 kDa, respectively. In our experience, many factors have been shown to have an important effect on the expression of gene products produced by filamentous fungi, and these are summarized in Table 5. Some of these will now be discussed in relation to the three examples listed.

In the case of culture pH, using the 67-kDa protein as an example (r-glucoamylase), there is substantial preference, in terms of production, for a postinduction pH of 5.0 over 5.5. This is illustrated in Figure 7. However, in the case of the 6-kDa protein (rhEGF, epidermal growth factor), there appears to be no such dependency, although the period required to accumulate maximum product postinduction is only approximately 20% of that required for glucoamylase, being only 7-8 hr in length. This is illustrated in Figure 8 (Smart and Sills, 1988). From our experience, induction time seems to be construct dependent, and this is further illustrated with reference to the 16-kDa protein, which takes approximately 20-30 hr to achieve maximum titres. This is illustrated in Figure 9.

Cell turnover during the postinduction phase is another factor that can seriously affect product expression. This is both due to nutritional problems as well as an apparent increase in the susceptibility of the cells to liquid shear forces after induction. The former point will be addressed first. Glucose is a potent repressor of the *alcA* promoter system, and even the very dilute concentration present at induction can severely reduce the level of expression possible. However, by using a combination of nonrepressing carbon sources such as alternative sugars, oils, and fatty acids, this problem has now been largely overcome. In relation to the problem

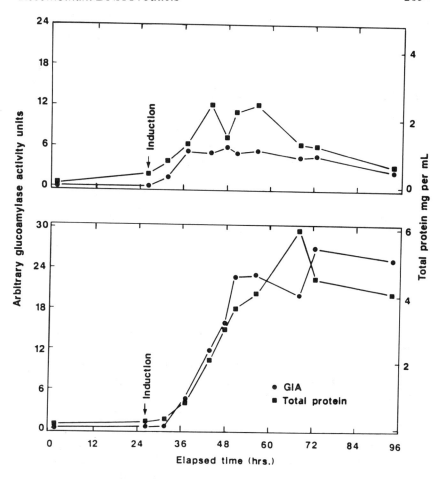

Upper culture pH 5.5

Lower culture pH 5.0

Figure 7 Production/secretion profiles for total protein and glucoamylase at different fermentation broth pHs.

of cellular turnover as a result of an increasing susceptibility to mechanical shear, by either altering the degree of turbulence in the vessel and/or the mixing patterns from radial to axial flow, increases in production yields have been shown to be possible. An example of this effect is illustrated in Figure 9.

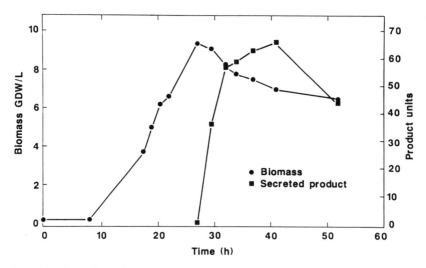

Figure 8 Secretion of rhEGF in a fermentation of *Aspergillus nidulans*.

As previously indicated, the degree of branching can be extremely important in relation to the yields of products obtained from the recombinant cultures of fungi, since it is believed that the sites of product biosynthesis occur at the tips of the mycelium. In relation to this, fermentation companies have selectively employed growth inhibitors to increase the level of branching in cultures of filamentous fungi, and in many cases this has had a significant effect on productivity.

This type of relationship has also been shown to occur in the case of recombinant cultures, and we have had some success in employing this methodology using "transition series" metals. This point is illustrated in Figure 10 using our *Aspergillus* expression system producing the 16-kDa protein, using an enriched ionic concentration of a transition salt at or around 0.3 mM. As indicated in Figure 10, this can produce quite dramatic effects, with yield improvements of 100% being possible (Smart and Sills, 1988).

Other morphological problems that can be encountered relate to the degree of dispersion of the fungal spores following inoculation into a rich production medium.

Our experience has been that the use of seed submerged cultures is not a prerequisite for recombinant pharmaceutical products, since the size of full-scale production vessels for many products is unlikely to exceed

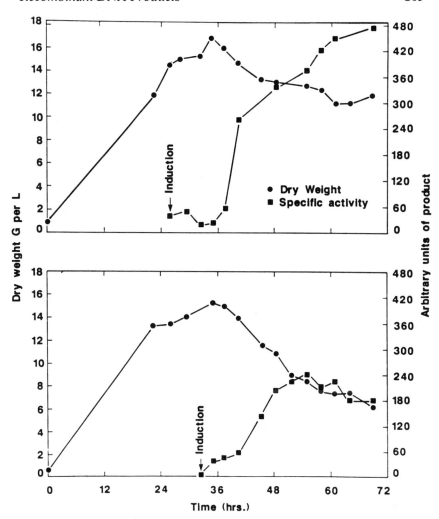

Figure 9 Effects of mechanical shear on the production of a recombinant protein in a genetically engineered strain of *Aspergillus nidulans*.

2000 l. In fact, using *A. nidulans*, our experience with submerged seed cultures suggests that these are not as productive, due to the relatively poor sporulation characteristics of the organism. Instead, using a seeding density of > 10^9 spores/ml of freshly prepared 4- to 5-day-old spores (harvested from agar trays), it is possible to achieve superior levels of expression

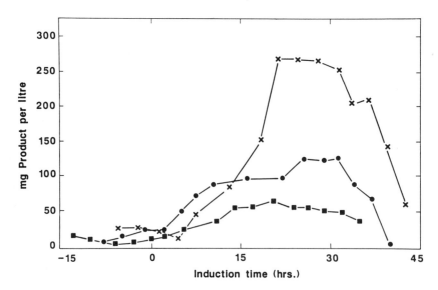

Figure 10 Increase in the production of a genetically engineered protein product as a result of a change in the organism's morphology under culture conditions.

by direct inoculation of the production vessel (Smart and Schwabe, 1988; Allelix unpublished data). However, in dealing with such large numbers of spores, it is sometimes difficult to ensure that full dispersion occurs, a condition that is necessary to prevent aberrations in the culture morphology. Like others who have exploited filamentous fungi, we have found that the use of polymers such as carbopol and alginate can aid in this area, although this is at a price of reducing the overall oxygen-transfer capability of the system (Elmayergi, 1975).

Like all recombinant systems, the selection of a suitable induction system is crucial for successful exploitation as an industrial producer of proteins. Although many alternatives now exist, the use of a simple industrial ketone has been shown to be particularly effective. However, as one might expect, several technical problems have arisen in using such a reagent, and many of these relate to the volatility of the substance and its removal from the fermentation broth. The following brief account is intended to illustrate how the removal of the inducer due to evaporation had a significant impact on the level of expression achieved using the *A. nidulans* system.

Variation in the production of recombinant product was correlated with the removal of inducer from the fermentation culture. In small-scale flask studies, maximum expression had been previously demonstrated at an inducer (ketone) concentration of 100 mM, although no studies relating to the evaporation of the ketone or the effective contact time for complete induction had been determined. Subsequent studies, however, indicated a rather slow removal from flask cultures, as detected by gas chromatography following CS_2 extraction (Watler and Smart, 1988; Allelix, unpublished data; Figure 11).

By comparison, removal from 10-l fermenter cultures was extremely rapid, as illustrated in Figure 12, and it was determined that the stripping coefficient for the solvent (inducer) was a linear function of the oxygen mass transfer coefficient, as given by the following equation:

$$K_a^{st} = 0.018K_L a - 0.027 \tag{6}$$

Later, model curves were fitted to this previously obtained experimental data, and from a subsequent mass balancing study, it was estimated that 4-5 hr of contact time was required to obtain full induction. This was estimated from the point at which no further losses occurred due to absorption or metabolism. This is illustrated in Figures 13 and 14. Since the generation of this data, physiological studies have confirmed this hypothesis (Bush and Sills, 1988; Allelix, unpublished data).

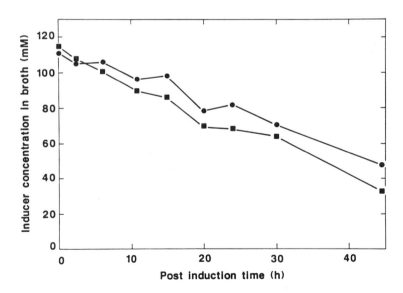

Figure 11 Plot showing the effect of inducer stripping from shake-flask cultures of *Aspergillus nidulans*.

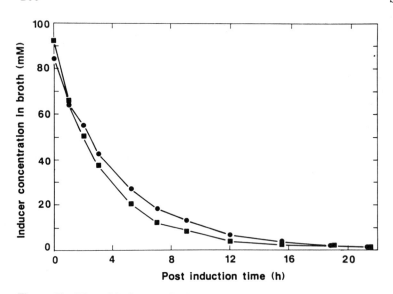

Figure 12   Plot of inducer stripping from fermenter broth samples of recombinant *Aspergillus nidulans*.

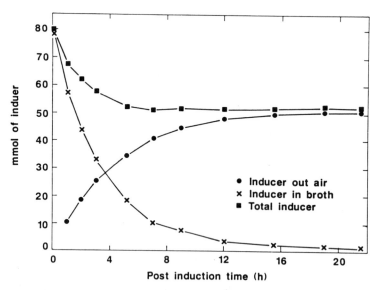

Figure 13   Inducer mass balance plot for fermenter cultures of *Aspergillus nidulans*.

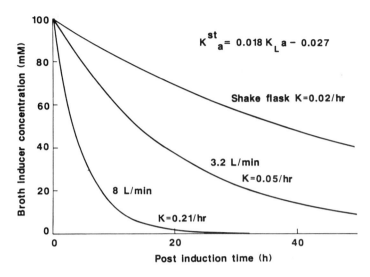

Figure 14 Stripping of chemical inducer from fermenter and shake-flask cultures.

As a result of these studies, several physical conditions were altered for the fermentation producing the 16-kDa protein, which would have some impact on the evaporation of the inducer. These included:

1. A reduction in the postinduction fermentation temperature by 5 °C.
2. A reduction in the postinduction aeration rate by 60%.
3. A reduction in the postinduction agitation rate by 40%.

The results are expressed in Table 6 as a percentage improvement over the control condition (Smart and Sills, 1988). In summary, all three conditions improved the production of recombinant product, although reducing the agitation condition seems to have the greatest impact on the yield. Intuitively, this is what might have been expected in view of the effect that agitation has on the oxygen mass transfer coefficient.

C. Expression Levels

To illustrate the potential of these expression systems, Figures 15, 16, and 17 show SDS gels of crude and partially purified samples for the molecules previously mentioned. In common with other expression hosts, the amount of protein produced seems to be product specific, although it is frequently the case that the target protein can be produced in quantities

Table 6   Effect of Reducing Ketone Inducer Stripping by Alteration of Postinduction Conditions

	Cell dry weight	Percentage improvement over control
Control	32.40	N/A
Reduction in temperature		
37°-28°C	31.64	80%
Reduction in aeration		
from 5 to 2 l/min	27.69	116%
Reduction in agitation		
from 1000 to 600 rpm	32.54	233%

N/A = Not applicable.

of > 50% of the total soluble protein synthesized. This may be observed in the case of both intracellular and extracellular proteins.

With reference to the examples cited, the r-glucoamylase shown in Figure 15 (a secreted protein) can be produced to a concentration of several grams per liter. Here lane 2 represents molecular weight markers; lane 3 is a sample of the untransformed *A. nidulans* culture; lanes 4 and 5 represent various stages of expression from the transformed cell line; and lane 1 is a partially purified sample of the protein.

In the case of Figure 16, which illustrates the production of the 16-kDa protein, this protein has been produced intracellularly to a concentration of many hundreds of milligrams per liter. Here, lane 4 represents molecular weight markers, lane 3 is a crude sample of culture broth, and lanes 1 and 2 represent dilute partially purified samples.

Figure 17 is an expression of secreted rhEGF at a concentration < 100 mg/l. Lanes 1 and 2 represent crude in-process samples at various stages of development; lanes 3 and 4 represent in-process samples from untransformed cells of *A. nidulans*, and lane 5 represents molecular weight markers.

IV. DOWNSTREAM PROCESSING

Once expression of the product has been achieved, the problem of recovering it from the culture begins. Although in many cases (using the Allelix *Aspergillus* expression system the product is expressed extracellularly, we have found that there are instances where intracellular expression may be beneficial or advantageous. In these cases an effective containable technique is required to disrupt the cells and release the product.

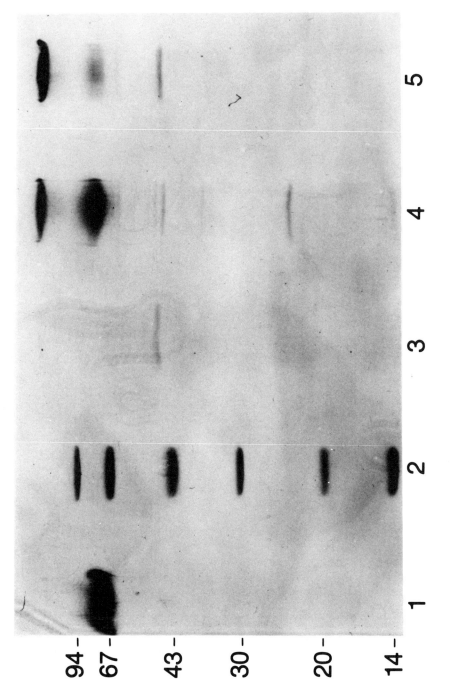

Figure 15 Expression of a 67-kDa protein by recombinant *Aspergillus nidulans.*

272

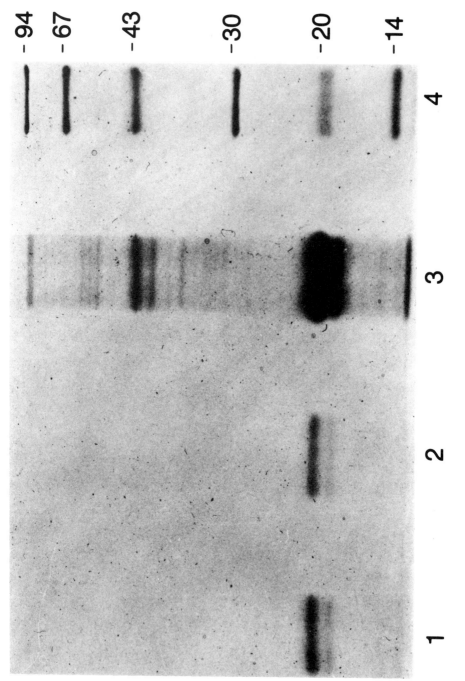

Figure 16 Expression of a 16-kDa protein by recombinant *Aspergillus nidulans*.

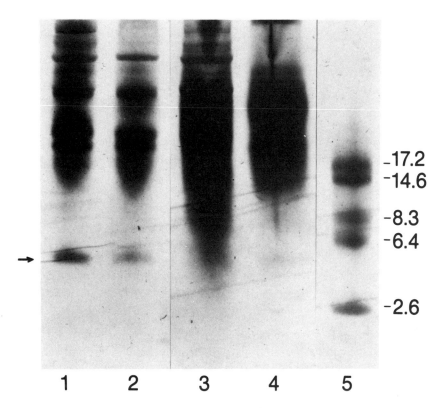

Figure 17 Expression of a 6-kDa protein by recombinant *Aspergillus nidulans*.

This is potentially quite difficult, since the technique used should not allow aerosols of the organism to escape into the surrounding environment nor should it provide too much energy to either deactivate the product or cause fragmentation of the fungal filaments to release structural proteins that might hinder the purification process.

From our experience we have found that these processes can be successfully accomplished using either a bead mill or a high-pressure homogenizer. (Lee et al., 1988; Allelix, unpublished data).

Figure 18 illustrates a schematic disruption profile for the release of alcohol dehydrogenase (ADH), an intracellular protein that serves as a model for other recombinant proteins. This data was generated using a laboratory-scale "dyno-mill" and 10-l laboratory-scale cultures of *A. nidulans*.

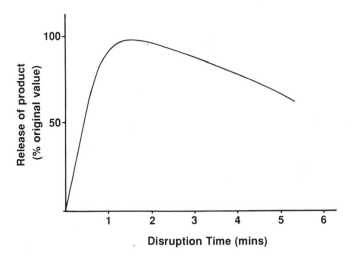

Figure 18 Schematic disruption profile for intracellular products synthesized by *Aspergillus nidulans*.

Disruption is affected within a period lasting between 1 and 2 min, provided the culture morphology is filamentous, and this is fairly consistent for cell mass loadings up to 35 g/l dry weight. Again, depending on the stability of the protein, in terms of both mechanical shear and/or thermal sensitivity, the product turns over if the processing time is extended. During such operations the material is usually incubated at approximately 10 °C and the mill agitator tip speed is run in the 10 m/s range. Additional factors, such as the inclusion of protease inhibitors or stabilizing agents, may be added to the processing buffer in the event of protease problems, although from our experience this has not been required.

For pelleted systems, the disruption time is required to be extended for up to 8 min to ensure that complete disruption of cells is produced, although this condition is rarely employed, owing to the fact that pelleted cultures are usually responsible for lower fermentative productivities and recovery yields are reduced due to processing damage.

Once the product is released into the processing buffer, microfiltration technology can be employed to clarify the concentrated solution. Use of the 0.2-μ Membralox ceramic cartridge* seems to perform well for this function, and a sample flux profile for a disrupted *A. nidulans* strain is

---

*Alcoa Separations Technology Division, 181 Thorn Hill Road, Warrendale, PA 15086-7527.

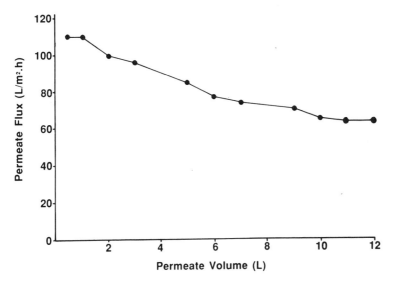

Figure 19 Decline in flux rate for disrupted cells of *Aspergillus nidulans* during microfiltration.

illustrated in Figure 19. This produces a cell-free solution within 2-3 h (depending on the transmembrane pressure and the filtration area used), which is suitable for further concentration by ultrafiltration or direct affinity absorbance.

In the case of secreted proteins, the primary separation process merely requires the cells to be removed before the product is concentrated using ultrafiltration technology. Depending on the local regulatory laws, this requires that the cells are either separated under contained conditions and then killed, or killed in the bioreactor before separation. Both of these routes pose problems, since the former requires a containable discharging basket-type centrifuge (or inverting filter centrifuge), because the containable disc-stack centrifuges used for bacteria are not compatible with fungal hyphae; whereas the latter procedure may damage the product because of its probable sensitivity to the deactivation procedure. Where possible we have adopted the former option due to the availability of a centrifuge with the characteristics mentioned.*

---

*Heinkel Filtering Systems, Inc., 520 Sharptown Road, Bridgeport, NJ 08014.

V. REGULATORY ISSUES

Regulatory issues play an extremely important role in determining the technical methodology employed in recombinant DNA technology and its use for the production of useful protein drugs. In employing recombinant filamentous fungi for such purposes, a specific set of issues need to be addressed if the processes used are to meet compliance.

A number of basic issues may be identified that need individual attention. These include:

1. Maintenance and propagation of the biological tissue
2. Special facility design considerations
3. Equipment design and selection
4. Effluent disposal
5. Occupational health and safety

A. Maintenance and Propagation of the Biological Tissue

For the manufacturing process, the strain to be used needs to be sourced from a master cell seed system, where fungal spores are suspended in a nutrient medium and stored in ampules at $-70\,°C$.

From this parent stock, the manufacturer's working cell bank (MWCB) is derived by recovery of the frozen material onto a rich nutrient agar previously formulated for maximum spore formation. This is usually carried out under "class 100" conditions to both prevent contamination of the production organism by aerial bacteria and to prevent unwanted release of the organism into the environment.

In terms of propagation, since one of the critical parameters is to maximize the number of growth centers, serial seed cultures are not frequently employed. Instead, direct inoculation of concentrated spore suspensions is the preferred route for recombinant pharmaceutical production, where production bioreactors are in the 500-3000 l range. Here the important thing is to develop a simple validatable protocol that is consistent and safe.

B. Facility Design

As a result of the stringent requirements for sterility and containment connected with process validation, and due to the fact that fungal spores are extremely resistant to sterilization, the movement of personnel and raw materials has to be strictly controlled in facilities where recombinant forms of fungal cells are to be cultivated (NIH Guidelines, 1987). In particular, for spore-bearing organisms such as fungi, Chapter 1 of the *Federal*

Register requires that "manufacturing processes should be performed in a completely separate building or in a portion of the building constructed in such a fashion as to be completely walled-off so as to prevent contamination of other areas, and should have an entrance to the area which is independent of the remainder of the rest of the building."

C. Equipment Design and Selection

In compliance with appendix K-III-A of the *Federal Register*, the cultures of viable fungal cells that have been modified by recombinant DNA technology have to be handled in a closed system that is designed to prevent the escape of viable material. This includes the bioreactors, processing lines, and tankage, in addition to the centrifuges, homogenizers, and UF systems.

A major problem arising from this directive relates to the contained separation of the fungal cells from the product-containing liquor, since the size of the hyphal filaments precludes the use of disc-stack centrifugation, which is used for separating bacterial and yeast cells. Systems that are traditionally used for separating fungal cells such as the vacuum rotary drum filter are difficult to contain, which makes validation a problem. One system that does solve this problem is the inverting filter centrifuge mentioned earlier, since it incorporates the attributes of both fast-action centrifuges and filter-press systems.

D. Effluent Disposal

This is monitored by the Environmental Protection Agency (EPA), and it is the company's responsibility to ensure compliance with the basic guidelines of the statutes, as well as any other local bylaw that may exist.

For fungal systems, the main consideration is to ensure that a validatable neutralization procedure is developed and implemented to prevent the propagation of in-process organisms that may escape as a result of aerosols or spillages. This should be given high priority, because of the potential of these organisms to survive outside the laboratory environment. Where possible, procedures should be developed during the "developmental" phase to prevent the formation of long-acting spores during "normal" vegatative exploitation of the cell line.

E. Occupational Health and Safety

Owing to the exemplary record of the fermentation industry with microbial cultures, the policy being adopted by OECD (Organization for Economic

Cooperation and Development) member countries recommends that regulations using traditionally safe microorganisms need not be altered where they have been modified by recombinant DNA technology. This is encouraging news for companies scaling-up recombinant fungal cultures, since their record with industrial enzyme and antibiotic production is well known, and therefore levels of containment exceeding BL-1 standards are probably not required.

Concerns about occupational safety are monitored by OSHA (Occupational Safety and Health Administration) and are usually coordinated internally by the quality assurance staff and biosafety team.

Major concerns about personal safety due to inhaled recombinant spores during inoculation procedures can be overcome by producing a validated transfer protocol, and this will be required to meet regulatory compliance.

Pyrogenicity of the new product is also an important consideration, so this will have to be thoroughly tested for, and a removal step must be incorporated into the purification scheme if this becomes a problem. To date, however, from our own experience, specific fungal pyrogens do not look as if they will pose a problem or add significantly to the total endotoxin load when compared to other expression hosts, and this is perceived as providing a possible production advantage for these systems for the large-scale manufacture of certain biologics.

VI. PROSPECTS

Although the use of recombinant fungal hosts is only in its infancy, especially for the production of protein pharmaceuticals, fungi possess the potential to make a significant impact, comparable to their use for the production of antibiotics and industrial enzymes. From the known systems, it would appear that fungi pose no obvious technical or regulatory disadvantages that might offset their apparent production advantages over other more established microbial systems.

ACKNOWLEDGMENTS

The author would like to thank David Gwynne, Michael Sills, Nigel Skipper, Peter Walter, David Schwabe, John Kabel, and Gordon Whitney for their useful comments and their contributions to some of the work discussed. He would also like to thank Kathleen O'Brien for typing the manuscript.

REFERENCES

Blanch, H.W. and Bhavaraju, S.M. (1976). Non-Newtonian broths: rheology and mass transfer. *Biotechnol. Bioeng. 18*:754-790.

Deindoerfer, F.H. and Gaden, E.L. (1955). Effects of liquid physical properties on oxygen transfer in penicillin fermentations. *J. Appl. Microbiol. 3*:253-257.

Einsele, A., Ristroph, D.L. and Humphrey, A. (1978). Mixing times and glucose uptake measured with a fluorometer. *Biotechnol. Bioeng. 20*:1487.

Elmayergi, H. (1975). Mechanisms of pellet formation of *Aspergillus niger* with an additive. *J. Ferment. Technol. 53*:722-729.

Gwynne, D.I. (1987). *The Bio/Technology Conference on Production Systems.* Grand Hyatt, New York, May 6, 1987.

Gwynne, D.I., Buxton, F.P., Williams, S.A., Garven, S., and Davies, R.W. (1987). Genetically engineered secretion of active human interferon and a bacterial endoglucanase from *Aspergillus nidulans. Bio/Technology 5*:713.

Margaritis, A. and Zajic, J.E. (1978). Mixing, mass transfer and scale-up of polysaccharide fermentations. *Biotechnol. Bioeng. 20*:939-1001.

Metzner, A.B. and Otto, R.E. (1957). Agitation of non-Newtonian fluids. *AIChE. J. 3*:3-10.

NIH Guidelines for Research Involving Recombinant DNA Molecules. (1987). August 24, 1987.

Smart, N.J. and Sills, M.A. (1988). Production of heterologous proteins using a recombinant *Aspergillus nidulans* expression system. Paper presented at the American Chemical Society Meeting, Microbial Technology Division, Los Angeles, CA, September, 1988.

Index

A

abaA 15
Absidia glauca 3
acetamidase (*see amdS*)
acetylCoa:deacetylcephalosporin C
   acetyltransferase 161
Achlya ambisexualis 3
acid phosphatase 56, 61, 68-70, 77,
   78
acid (*see* aspartic protease and
   protease)
Acremonium chrysogenum 150-151,
   156-161, 165-168, 172, 175-
   177, 182-185, 252
   mutants impaired in cephalsoporin
     C synthesis 165-166
   strains with improved productivity
     167, 176
acrolein 62
actA 15
actin 14

acuD (isocitrate lyase) 15, 179
ACV [δ(L-α-aminoadipyl)-L-
   cysteinyl-D-valine] 150, 152,
   156, 162, 163, 168,
ACV cyclase (*see also pcbC*) 150-
   152, 163, 165, 167, 168, 170-
   176, 180, 181, 184, 185
   essential amino acid residues 171-
     172
   evolution/phylogenetic
     relationships 169-172
ACV synthetase (*see also pcbAB*)
   150, 152, 185
acylCoA:IPN acyltransferase 151,
   153, 154, 185
   amidolyase subunit 153
   acylCoA:6-APA acyltransferase
     subunit 153-156, 162, 163
adeZ 218
adhA 16, 35, 38, 40
ADH I 50, 62-64, 271
adh3 8, 15

F

G